MW01147081

The MIT Press Essential Knowledge Series

A complete list of books in this series can be found online at https://mitpress.mit.edu/books/series/mit-press-essential-knowledge-series.

PHENOLOGY

THERESA M. CRIMMINS

The MIT Press | Cambridge, Massachusetts | London, England

The MIT Press would like to thank the anonymous peer reviewers who provided comments on drafts of this book. The generous work of academic experts is essential for establishing the authority and quality of our publications. We acknowledge with gratitude the contributions of these otherwise uncredited readers.

This book was set in Chaparral Pro by New Best-set Typesetters Ltd. Printed and bound in the United States of America.

Library of Congress Cataloging-in-Publication Data

Names: Crimmins, Theresa M., author.
Title: Phenology / Theresa M. Crimmins.
Description: Cambridge, Massachusetts : The MIT Press, [2025] | Series: The MIT Press essential knowledge series | Includes bibliographical references and index.
Identifiers: LCCN 2024019250 (print) | LCCN 2024019251 (ebook) | ISBN 9780262551052 (paperback) | ISBN 9780262381543 (pdf) | ISBN 9780262381550 (epub)
Subjects: LCSH: Phenology.
Classification: LCC QH544 .C75 2025 (print) | LCC QH544 (ebook) | DDC 578.4/2—dc23/eng/20241108
LC record available at https://lccn.loc.gov/2024019250
LC ebook record available at https://lccn.loc.gov/2024019251

10 9 8 7 6 5 4 3 2 1

CONTENTS

SERIES FOREWORD

The MIT Press Essential Knowledge series offers accessible, concise, beautifully produced pocket-size books on topics of current interest. Written by leading thinkers, the books in this series deliver expert overviews of subjects that range from the cultural and the historical to the scientific and the technical.

In today's era of instant information gratification, we have ready access to opinions, rationalizations, and superficial descriptions. Much harder to come by is the foundational knowledge that informs a principled understanding of the world. Essential Knowledge books fill that need. Synthesizing specialized subject matter for nonspecialists and engaging critical topics through fundamentals, each of these compact volumes offers readers a point of access to complex ideas.

PREFACE

I have been working on and carrying out research in the field of phenology for nearly twenty years, and for the majority of those years, my family members have repeatedly asked me, "Now what is it you do again?" Who can blame them? Phenology is a vague and arcane term, conjuring either images of skull morphology or . . . nothing at all.

I have for years craved a reference they could consult to better understand phenology and why it matters. While academic literature characterizing how phenology is changing and the impacts we are experiencing abounds, resources that succinctly summarize this information for a more general audience are more limited. I hope that with this book, I am addressing that need.

1

WHAT IS PHENOLOGY?

Bill reaches up and scratches his head, adjusts his glasses, and points out a leaf-footed cactus bug eating a saguaro cactus flower bud. A geologist by training, Bill Peachey has been tracking flowering in saguaro cacti east of Tucson, Arizona, United States, for close to thirty years. And when I say "tracking," I mean he documents every single flower bud that appears on each of the 406 arms of every one of the 141 individual cacti in a 2.5-acre plot nearly every day of the late spring flowering period, which extends from May to July. Today, Bill and I are visiting his plot to check on how many blooms have developed since his last visit two days ago.

Bill's interest in saguaro flowering was originally sparked in the late 1990s when he was researching the Mexican long-tongued bat. Found only in the Sonoran Desert of the southwestern United States and northwestern

Mexico, saguaros are a key food source for the Mexican long-tongued bats. The distinctive columnar cacti provide shelter, food, and water to over a hundred other native animals in the region as well, including other bats, white-winged doves, sparrows, orioles, desert tortoises, foxes, and coyotes. The original study focusing on the bats finished up in 1997 and revealed novel findings about what plants the bats use as nectar sources. But Bill's focus on the saguaros proved to be a worthwhile endeavor: the more Bill talked with local naturalists and experts, the more he learned that long-held beliefs about the cacti were based on limited information or even wrong. His efforts to carefully document flowering and fruiting activity were advancing basic knowledge about this unique, charismatic, and important cactus. And so Bill kept tracking the cacti and their flowering patterns.

Over the twenty years he has been carefully watching them, Bill has amassed over 136,000 observations of saguaro blooms. He has observed clear variation from one year to the next as well as interesting changes within the plot. In 2005, flower abundance was massive; Bill counted over 14,000 flowers. And 2021 was another remarkable year, with 12,000 blooms—almost twice the number in a "typical" year. Then in 2022, there were only about 50 blooms in total. Further, between 2004 and 2013, the start of the flowering period drifted later, and the period of flowering shrank.[1] Why? Is the timing of flowering in these important cacti changing, as things are in other plants in the region? And if it is, what are the implications, not only for the saguaro, but for animals like white-winged doves, desert tortoises, and foxes that depend on the flowers and fruit for food, and the Sonoran Desert as a whole?

Bill's focus—and his questions about what cues flowering in saguaros, why flowering timing and numbers vary so dramatically from one year to the next, and whether there are long-term trends at play—is the heart of the study of phenology.

Defining Phenology

How do you know when spring has arrived? I grew up in Michigan, United States. My recollections of the start of spring include the reappearance of warm, sunny days that meant I could shed my bulky winter parka and the piles of dirty snow in the corners of parking lots shrinking down to icy lumps. I also recall showy white dogwood blooms gracing otherwise bare branches in the surrounding oak-pine forest, the earthy smell of spring rains falling on decaying leaves and soil, and the reemergence of woodpeckers' staccato drumming. I now live in southern Arizona, in the Sonoran Desert. I know spring has arrived in my desert home when the sun is inching above the horizon before my alarm goes off, lizards begin to scurry around my yard again, and flowering blue dicks—short herbaceous plants with clusters of delicate purple flowers borne on long, slender stems—grace the low desert.

Notice that these definitions of spring include both changes in environmental conditions—increasing daylength and temperature—and developmental changes in plants and animals—the appearance of flowers and start of activity in lizards. We experience seasons—changes in the nonliving, environmental conditions including daylength and temperature—because of the earth's tilt and revolution around the sun. The earth rotates on an axis that is not perpendicular to the plane of revolution around the sun. Consequently, for half of the year, the Northern Hemisphere is angled toward the sun, and experiences longer days and warmer temperatures, while the Southern Hemisphere is angled away, and experiences shorter days and cooler temperatures. For the other half of the year, this pattern reverses. Plants and animals respond to these changes, increasing their activity as conditions become favorable, and reducing their activity when conditions become too hot or cold, too wet or dry, or otherwise unsuitable.

If you pay attention to seasonal changes in your environment, you most certainly have noticed that seasonal events don't happen on the same day every year. Rather, you might see dramatic variation in when events like flowering begin from one year to the next. In 2021, I saw few springtime cactus blooms in my neighborhood in Tucson, and the ones I did see flowered in late May. In sharp contrast, spring 2023 brought us an outrageous abundance of cactus blooms that graced our yards and roadsides from February until mid-June. This year-to-year variation occurs because of global weather patterns along with the influence they have on plants and animals.

Regardless of where you live on the earth, seasonal events in plants and animals like leaf out or flowering, first songs, egg hatch, and migration given definition to the year, in response to fluctuations in daylength, daily temperatures, and rainfall patterns. When these seasonal shifts occur is termed "phenology," in what I like to think of as an old-fashioned, awkward term for phenomena that are so familiar to us, they sometimes fade into the background of our awareness. Though often confused with "phrenology," another old-fashioned word that refers to the pseudoscience of using the shape and dimensions of one's skull to predict mental capacity, the two terms could not be more different! *Phenology* is a real, legitimate aspect of biology and scientific study.

How Do Scientists Define Phenology?

In short, "phenology" refers to when recurring seasonal events occur; it's all about timing. The focus is typically events that occur on an annual basis in plants' and animals' lives, such as leaf out, flowering, and fruiting, and emergence from hibernation, mating, rearing and fledging young, migration, and reentry into hibernation. Phenology also encompasses less visible seasonal events, including the commencement of root growth as well as events in other groups of organisms including fungi and microbes, such as when mushrooms pop up and release their spores. Recurring seasonal events in nonliving things are sometimes lumped in with phenology too. The dates of ice on in the fall and ice breakup in the spring on several major lakes and rivers have been carefully tracked for many decades by local organizations. Authoritative definitions, however, typically confine "phenology" to seasonal events in living things, and classify seasonal events in nonliving things like lakes and rivers as "seasonality."[2]

The term "phenology" was first used in 1849 by Belgian botanist Charles Morren and is based on the root "phaino," as is the same for the word "phenomenon."[3] Organisms including plants, animals, and fungi are sensitive and responsive to their immediate surroundings; changes in daylength, temperature, and moisture as well as what neighboring critters are doing govern when animals and plants begin and end their seasonal activities. For example, a cool summer can delay fruit ripening, a dry spring can lead to earlier senescence—the breakdown of tissue at the end of the growing season—among short-lived spring plants, and especially warm temperatures often result in

“Phenology” refers to when recurring seasonal events in plants and animals occur.

earlier spring and summer activity among both plants and animals.

The term "phenology" strictly refers to the study of ("-ology") recurring biological phenomena ("pheno") and the relationship with environmental conditions. As such, it doesn't make sense to use the term "phenology" to refer to the timing of seasonal activity in an organism as the organism's phenology. Yet it is common practice within the scientific community to do this very thing, and few clear synonyms exist. Given this, I use "the timing of seasonal activity" with "phenology" interchangeably throughout this book.

Humans Have Observed Phenology for Millennia

Since humans' first appearance on the planet, our existence has been inextricably tied to seasonal changes in plants and animals. Lacking grocery stores or even refrigeration, early humans had to rely on locally available plants and animals as sources of food. An awareness of the seasonal growth and migration patterns was critical for survival. Cave paintings in Europe dating to fifteen to forty thousand years ago document the mating seasons of birds, bison, deer, fish, and horses using a lunar calendar, demonstrating that early humans recognized—and tracked—the cyclic nature of biological events.

Cave paintings in Europe dating to fifteen to forty thousand years ago document the mating seasons of birds, bison, and fish, demonstrating that early humans recognized the cyclic nature of biological events.

Cues such as daylength and the position of the stars have long been used to indicate the optimal time to harvest resources for food, materials, and medicines in many cultures. For instance, Indigenous communities across Canada use singing in Swainson's thrush, or "salmonberry bird," as an indicator that salmonberries—related to raspberries—are ripe and ready to pick.[4] Hundreds of examples like this have been documented. Many Indigenous peoples continue to integrate their understanding of phenology into resource management, subsistence lifeways, and ceremony.[5] The Nlaka'pamux people of interior British Columbia, for example, know to dig roots of the bitter root plant, a critical winter food resource, and celebrate the arrival of spring with "first roots" ceremonies when saskatoon plants begin to flower.[6] Similarly, the Lenape people of the woodlands of the northeastern United States and southeastern Canada use flowering in this early spring shrub to signal the massive run of shad up the Hudson River, thereby signaling the time to fish.[7] Accordingly, this plant is known as "shadbush" in this part of the world. These communities' deep understanding of the relationships between seasonal changes in the environment and natural events has shaped the timing of harvesting and gathering, agricultural activities, resource management, and ceremonial activities for tens of thousands of years. Some fantastic resources that expand on this topic are listed in the further reading section below.

Formal recordkeeping of seasonal events dates back several thousand years in many parts of the world. In China, a calendar based on twenty-four periods was established in the eleventh century BCE, and evidence of routine observations of phenological events date to nearly four thousand years ago.[8] The practice spread to ancient Greece, where phenology observations carved on limestone tablets date to the fifth century BCE.[9] In Japan, the timing of flowering in cherry trees has been reconstructed from diaries, chronicles, and other formal records stretching back to the early eighth century CE.[10] In addition, calendars based on phenology have been used to shape agricultural practices worldwide for thousands of years.

Seasonal Events Are Shaped by Environmental Conditions

The timing of seasonal events is the result of recent local environmental conditions. Though seasonal events often occur at approximately the same time each year—for example, red maples flower in late February and early March in southern states in the United States, and late March to early May in northeastern states—the precise timing varies because of the weather conditions in a particular year. If we can identify the particular weather conditions that are associated with an event, we can then better anticipate when that event will happen simply by watching the weather. Recognizing the clear ties between local conditions and plant and animal activity, my friend Jeff in Arizona recently told me he knows that when he starts to hear the loud buzzing of cicadas—large, somewhat creepy-looking insects with wide-set eyes and large wings—"it's going to get flippin' hot out."

The notion that weather conditions directly influence the timing of seasonal events in plants and animals was famously documented by Carolus Linnaeus, the Swedish botanist known for establishing the system biologists use for naming and categorizing organisms. On a notably cold spring expedition in 1741, Linnaeus apparently remarked that "spring should be measured according to climate and temperature rather than by the calendar," demonstrating his recognition of year-to-year variability in both weather and plant response.[11] In perhaps the first study of the kind, Linnaeus recorded both the dates of leafing, flowering, fruiting, and leaf fall for many species in combination with climatological variables at eighteen estates in Sweden. This study was unique in that daily weather and plant response were recorded at the same locations, and also because these observations were collected across multiple locations. This enabled Linnaeus to investigate relationships between local weather conditions and plant response as well as differences among the sites. Linnaeus documented his methods in the well-known *Philosophia*

Botanica in 1751 so that others might adopt his approach and thereby advance the collective understanding of the environmental variables controlling plant activity.

A mechanistic approach to understanding phenology and the associated environmental conditions flourished throughout the scientific revolution and age of enlightenment, with an emphasis on the scientific method and reason. Around the same time that Linnaeus was observing flowering across Sweden, a French entomologist and writer, René de Réaumur, took note of the relationship between plant leaf and bloom dates and the temperatures in the preceding months. He formalized a method of quantifying "warmth" within a period preceding an event such as leaf out, referring to these thermal units as growing degree days, in a seminal publication in 1735 that is still referenced today.[12] Several others through the nineteenth and twentieth centuries refined this thermal sum concept. Today, many techniques for calculating accumulated thermal units are widely used to predict seasonal activity in plants and insects.

Approximately a century later, Adolphe Quetelet, a professor of mathematics at the Athenée de Bruxelles as well as physics and astronomy at the Musée de Sciences et des Lettres in Belgium, reported that plants flower when exposed to a specific quantity of warmth. Based on his observations of flowering time in hundreds of plants collected at the Brussels Observatory between 1833 and 1852, he published his "Law of Flowering Plants," which stated that plants flower after exposure to a specific quantity of heat.[13] Quetelet's method differed from that of de Réaumur's by using degrees of Celsius squared to calculate thermal time, which importantly emphasizes springtime "warm spells"—notably warm conditions that span multiple days and appear to have outsized influence on triggering plant growth. His primary motivation for this work was to characterize the climate of a region. Quetelet argued that the climate of a region could be better understood by characterizing "periodical phenomena" such as flowering and published a set of observation protocols to enable others to establish such relationships in other climates.

Morren, a former student of Quetelet's and accomplished scientist, appreciated the significance of tracking seasonal phenomena, but resisted Quetelet's narrow focus on the climatic conditions associated with flowering. He saw untapped potential in better understanding plants' relationships to not only climate but also each other. Finding inspiration in Linnaeus's work, Morren argued for increased rigor in the species selected for monitoring and the specifics of observations, recommending that observations encompass not just flowering onset but duration and frequency too. Quetelet and Morren debated approaches, areas of focus, and terminology for this emerging field for over a decade in the mid-1800s, and both left indelible fingerprints on the field. Quetelet's interest in the conditions associated with phenological events has become a major focus of the field, and Morren's proposed term "phenology" has persisted.

In the early 1900s, Andrew Delmar Hopkins, a US entomologist, added another important dimension to the field of phenology by extending the relationships between phenological events and temperature across space. In his extensively applied bioclimatic law, Hopkins proposed that the timing of seasonal events is delayed by four days with every degree north in latitude, for every five degrees west in longitude, and four hundred feet up in elevation.[14] Hopkins used these laws to advise farmers on when to sow winter wheat to avoid pests and frosts.

Phenology Shapes Local Environments

Phenology—when plants and animals are active—also influences where species are found and how ecosystems function. For example, migrating animals transport other plants and animals to new locations. Plant pollen and seeds can hitch a ride on animal fur, feathers, or feet as well as inside the gut of the migrating creature, then deposited in a new location—along with nutritious fertilizer—when the animal defecates. Insect, amphibian, or mollusk eggs can likewise attach to the feet or other parts of larger migrating birds, mammals, or bats. Parasites and pathogens can be introduced to new areas with the arrival of migrants too. Moreover, migrating animals move nutrients and energy from one area by consuming plants and animals and defecating along their journeys and serve as a food source for other creatures. If the timing of animal migration shifts, any or all of these phenomena can be affected. Similarly, the timing of plant growth shapes which plants populate a community. When plants adjust their leaf out earlier or later in the year in response to changing conditions, they can create opportunities for new—and sometimes problematic—plant species to move in.

Plant and animal growth and activity also shapes nonliving aspects of the environment. For instance, the appearance of leaves in the spring changes the amount of radiation coming from the sun that reaches the ground as well as the "roughness" of the land surface, thereby affecting wind currents and local air circulation patterns. In addition, the onset of seasonal growth in plants increases the uptake of carbon dioxide by plants. This, in turn, influences the timing of seasonal events occurring later in the season.

Phenology Is a Conspicuous Indicator of Environmental Change

My colleague and friend Melanie recently remarked that her May birthday, which used to fall during an abundance of yellow blooms on the landscape, is now a "pink birthday."

The pink blooms that presently coincide with her birthday are those of the ironwood tree, a gorgeous tree common in southern Arizona. The yellow blooms that historically graced her birthday are those of the palo verde tree, another abundant tree of the Sonoran Desert. I hear reports of changes such as these on a regular basis. My experience is that as soon as I define the term "phenology," my conversation partner is inspired to share their own experience of change in a noteworthy seasonal event.

Because the timing of seasonal events in many species is highly sensitive to conditions such as temperature, events such as leaf out and emergence from hibernation are demonstrating clear changes as global temperatures increase. In 2007, phenology was singled out as "the simplest process ... to track changes in the ecology of species in response to climate change" by the Intergovernmental Panel on Climate Change (IPCC), an international body tasked with assessing the science related to climate change.[15] Evidence of changes in the timing of events is piling up from locations around the globe, from traditional data sources such as long-term observations as well as surprising origins such as wedding photos and video footage of sporting events.

Even though the term "phenology" remains unfamiliar to most people, the topic has gained significant attention in recent years. In the next chapter, we will dig deeper into how phenology is changing, as changes are not consistent among species or regions. Chapter 3 covers the sources of information that scientists have used to determine how plants and animals are shifting their phenologies, including traditional observational approaches and surprising sources of information like weather radar. Chapter 4 articulates why changing phenology matters to us in our everyday lives, and chapter 5 details how a deeper understanding of the environmental conditions associated with seasonal events can be used in a predictive way.

Since awareness of climate change arrived on the scene in the 1970s and 1980s, the call for action to address the root causes and consequences has grown increasingly louder and more insistent. Scientists and conservation organizations have pressed for increased monitoring to better inform decision-making and policy, and ultimately curtail species loss. Phenology is repeatedly identified as a critical measure to track, as it is highly visible, familiar, and accessible to all. Just about everyone understands "blooming" and "mating," and anyone can document these phenomena—no formal training or specialized equipment necessary. In chapter 6, I invite you to weave a practice of observing the seasonal cycles of plants and animals into your own life to contribute to science as well as soothe your soul.

Though I aim to share examples of phenology from around the globe, the content of this book is biased toward North America, both because of the preponderance of studies that have taken place here and because it is where I have lived my own experience.

HOW IS PHENOLOGY CHANGING?

Winter and spring 2023 were very warm in much of the eastern United States, and springtime biological activity started much earlier than normal. The news media took note, with dozens of outlets trumpeting the "earliest spring on record," a premature beginning to the allergy season, and an early start to flowering in the iconic cherry trees of the mid-Atlantic region. Even the popular US television show *Saturday Night Live* gave it a nod, with an opening skit in the April 15, 2023, episode featuring early sightings of stereotypical characters who populate Central Park when the weather is nice.[1] Actors in the skit acknowledged that New York City had already reached 90°F, three months ahead of schedule. Many of the clearest changes in phenology around the globe are occurring in spring. While we might appreciate the break from bulky winter clothing and scraping frost from our windshields brought on by an early arrival of spring, changing climate conditions and progressively earlier springtime conditions create challenges for both plants and animals, with repercussions that ripple throughout ecosystems and extend to humans.

While we can't wholly blame the warmer and earlier spring 2023 on climate change, warmer years are definitely a common occurrence these days. Recent and rapid changes in the earth's climate increase the chances of earlier springtime activity in any given year. The primary culprit for the increases in global temperatures in recent decades is the addition of greenhouse gases into the atmosphere, primarily through the burning of fossil fuels. These gases, mainly methane and carbon dioxide, trap incoming solar radiation as heat. Since preindustrial times, global atmospheric carbon dioxide concentrations have increased by more than 50 percent, from around 278 parts per million (ppm) in the 1850s to around 417 ppm in 2022. In response, the earth's mean temperature has increased by 1.1°C.[2]

Global temperatures continue to creep upward; globally, 2015 to 2023 were the eight hottest years on record as of this writing.[3] We are also experiencing more extreme weather events: summer heat waves are more frequent, longer, and more intense; extreme rainfall events are more frequent; and tropical cyclones are more common.

Increases in temperatures have not been even around the globe: warming is more rapid and intense at higher latitudes as well as in the winter season. Warmer and shorter winters result in reduced winter ice cover along with earlier springtime ice breakup and meltwater flows. Since the beginning of the twentieth century, ice breakup in the northeastern United States has advanced by 1 to over 3 weeks. Over the past 165 years, ice duration has decreased by 31 days in the Northern Hemisphere.[4]

It is important to put these comparatively recent, notable changes in our climate and earth systems into a longer context. Reconstructions of the earth's climate made using data from tree rings, ice cores, and lake sediments reveal dramatic fluctuations in the earth's climate. The period spanning approximately 1300 to 1850 CE, referred to as the "Little Ice Age," was characterized by glacial expansion on several continents and resulted in comparatively cool global temperatures. Stretching further back, global temperatures were comparably very warm around 6,500 years ago. Such fluctuations demonstrate that the earth's climate was not historically static. Yet the most recent IPCC report, authored by worldwide experts on climate and related subjects, indicates that the rate of temperature increase that we have witnessed in recent years is "unprecedented" in the last 2,000 years and "unequivocally" the result of human activity.[5]

How Are Plants and Animals Responding to the Changing Climate?

In response to changing conditions, plants and animals have three options: move to where conditions are more suitable, adapt to the new conditions, or give up and go extinct. A major way that species adapt to novel conditions is by shifting the timing of their activity, undergoing transitions like leaf out or egg hatch earlier or later in the year. Around the globe, evidence of species shifting their phenology in response to rapidly changing conditions is piling up.

When all the results from around the world are toted up, we find that the timing of spring events like leaf out, flowering, and bird arrivals has advanced by approximately four days since the 1980s, though the numbers vary considerably by region, taxonomic group, and season.[6] Springtime events generally show clearer and more dramatic changes than autumn ones. Plants display the greatest advancements in the timing of their activity, as do primary consumers—animals that eat plants. Warm-blooded animals higher on the food chain, such as birds and mammals, show smaller shifts than cold-blooded animals like insects and amphibians.[7] Further, organisms active earlier in the season evidence greater rates of advancement than those active later in the season.[8]

Around the globe, evidence of species shifting their phenology in response to rapidly changing conditions is piling up.

Around the Globe, Plants Are Shifting Leaf Out, Flowering, and Fruit Ripening

In ecosystems all around the world, plants are responding to changing climate conditions by shifting their activity in all seasons. The magnitude of these changes, however, varies dramatically by season, taxonomic group—amphibians seem to be shifting activity more than mammals, for example—and region. The biggest shifts in springtime activity are occurring in Asia, followed by Europe, though our ability to state this with confidence is hampered by limited information from the Southern Hemisphere. On each of these continents, the timing of leaf out in common overstory trees has shifted from nine and fifteen days earlier since the early 1980s.[9] Changes in the timing of leaf out are less dramatic in the United States; since the early 1980s, spring leaf out has advanced by about a week.[10] We have the clearest understanding of how plants are adjusting the timing of their activity in temperate ecosystems in Europe and the United States; our knowledge of ecosystems on other continents is sparser. Even so, some long-term observational datasets do exist in semiarid, arid, and arctic ecosystems, and they all reveal similar patterns, with general trends of advancing springtime phenology and variability from one species to the next.[11]

A fantastic example of how overstory deciduous trees and understory herbaceous plants from temperate systems have adjusted the timing of spring activity since preindustrial times hails from New York. In the early 1800s, instructors at secondary schools across the state regularly documented when they observed meteorologic conditions as well as plant and animal activity. They continued this effort until the Civil War. These data languished, all but forgotten in dusty annals, until Dr. Conrad Vispo, cofounder of the Farmscape Ecology Program in Hawthorne Valley, discovered them in 2014. Vispo immediately began to digitize the old data, appreciating the gold mine these historical records offered. Yet he recognized that they had little scientific value without a modern context. So he searched "phenology in New York" online and discovered the New York Phenology Project, a statewide phenology monitoring effort that had been initiated in 2012 by Dr. Kerissa Fuccillo Battle. The two scientists were astounded when they realized the similarity between the two efforts. They began a collaboration on the spot. When the two datasets were compared, the findings were eye-popping. Tulip trees now flower twenty-seven days earlier than in the 1800s, and common milkweed, a plant that is critically important to endangered monarch butterflies, now flowers thirteen days earlier. Overall, plants in the study leaf out nineteen days sooner than they did in the early 1800s.[12]

While the results of the New York study are impressive, they are not novel. Dozens of studies have shown impressive shifts in springtime plant and animal activity in the temperate forests of the United States. In West Virginia, two spring wildflowers, cutleaf toothwort and yellow trout lily, now flower approximately six days earlier than 100 years ago.[13] In Mohonk, New York, hepatica, bloodroot, and trout lily now flower over a week earlier than in the 1930s.[14] In Concord, Massachusetts, yellow wood sorrel now flowers thirty-two days earlier than 150 years ago.[15] This pattern extends to Europe. In Switzerland, horse chestnut has shifted leaf out by eleven days since the early 1800s, and cherry trees have advanced their flowering time by thirty days over the past 130 years.[16]

But not all plants are shifting their springtime activity earlier in the year. Some are exhibiting the opposite response, delaying their springtime activity. For example, in North Dakota, nannyberry, smooth sumac, and black walnut now flower approximately ten days later than in the first half of the 1900s.[17] And many other species seem not to care much about changing conditions and persist in undergoing springtime activity at the same time that they have in decades past. This is especially true in the Southern Hemisphere. Approximately 70 percent of the species evaluated in a comprehensive study in Australia and New Zealand showed no evidence of changes in phenology.[18]

A handful of generalizations is starting to emerge from the hundreds of studies of plant phenology from around the globe. In general, plants active the earliest in the spring show the greatest changes in the timing of their activity. In addition, annual plants, which complete their entire life cycle—from germinating to producing mature fruits or seeds—within a single year, generally evidence larger advancements than their longer-lived counterparts. Further, shifts in the timing of leaf out and flowering appear to be greater for wind-pollinated plants than those that are pollinated by insects. Wind-pollinated plants produce abundant pollen grains that are small and lightweight enough to be carried on air currents from one plant to another. These plants are the ones responsible for causing seasonal allergies, as the pollen grains produced by these plants are small enough that we breathe them into our lungs. In contrast, the pollen grains produced by insect-pollinated plants are comparatively large and often sticky or spiky, enabling them to hitch a ride on the insects visiting the plants for nectar. The trend toward earlier flowering in wind-pollinated plants is not good news for those of us suffering from seasonal allergies; this means an earlier start to the allergy season! Shifts toward earlier activity are often greater for plants introduced to an environment than those exhibited by native species. And finally, plants at higher elevations and latitudes frequently show greater changes in their timing than their lower-elevation and latitude counterparts.

For many of us, autumn is a favorite season, bringing cool, crisp days, brilliant gold, crimson, and orange leaves, and pumpkin spice everything. Around the globe, however, the natural events that signal our autumn experience, like leaf color change and leaf fall, are changing as well.

In general, autumn events are drifting later in the year than in decades past, though patterns in autumn are not as clear or strong as those emerging for spring. Autumn phenophases—life cycle stages such as leaf color change and leaf drop—appear to be shaped not only by temperatures, beginning when temperatures start to cool in the late summer, but also by daylength and available moisture. In general, when temperatures are warmer, leaf senescence—when the chlorophyll that is responsible for giving leaves their green color breaks down—is delayed. Moreover, the reveal of the brilliant yellows and oranges that were there all along but masked by green as well as the generation of the anthocyanins responsible for dazzling red in some leaves are both delayed. Yet midsummer drought can cause leaf color change to occur earlier in the season or even cause leaves to drop with no coloration at all. The opposite is true for fruit ripening, though. Following suit with leaf out and flowering, fruit ripening is generally occurring earlier in the year now than in the past in response to warmer temperatures.

Animals Are Shifting Their Phenologies Too

One mid-June morning in 2023, I was delighted to wake up to a family of Gambel's quail in my backyard. I counted ten tiny fluffs, scurrying along behind their watchful mother, who made persistent cooing sounds to direct their movement. Smaller than a chicken's egg, the downy little things scuttled along in random directions while both bird parents strained to herd them to safety.

Mid-June seemed late in the season for quail babies, but I'm no bird expert, so I reached out to someone who is. Dr. Jennifer Gee, director of the University of California James Reserve, and her colleagues corroborated my anecdotal observation; they had also observed later-than-usual chick arrival that spring. Winter and spring were quite cool and wet in the southwestern United States, and plants flowered later than usual across the region. Cooler and wetter winters typically lead to later springtime activity in plants and animals in this region. Gee and her team reasoned that the cooler temperatures and slower response among the plants led the birds to similarly delay their springtime activity.

Scientists and naturalists worldwide have been captivated by birds for centuries. The meticulous checklists that many birders maintain are a rich source of information on avian comings and goings, and reveal that globally, birds have advanced the start to their breeding seasons at a rate of two to three days per decade since the early nineteenth century—a shift totaling forty to sixty days![19] These records also reveal that some migratory birds are shifting the timing of their spring arrival more than others, and this is affecting their survival and thus population numbers. Short-distance migrants are birds that migrate tens to hundreds of miles between their summer breeding and wintering grounds, typically staying on the same continent. In North America, this includes American robins, waxwings, red-winged blackbirds, and several other common species. Long-distance migrants cover distances often numbering in the thousands of miles and frequently include a move between two continents. Most birds we see here in North America are long-distance migrants, spending their summer months in North America and then flying south to Central or South America for the winter.

Across the Northern Hemisphere, short-distance migrants are shifting the timing of their spring arrival more than long-distance migrants, and this has everything to do with what information the birds use to decide when to start their movement northward. Short-distance migrants pay attention to the temperature of the surrounding environment, departing for their summer breeding grounds once temperatures seem warm enough to be "spring." In contrast, long-distance migrants rely on daylength—the number of hours of daylight in a twenty-four-hour period—to know when it's time to go.

The different strategies short- and long-distance migrants employ for cuing migration were magnified in 2012, a year with a very warm spring in the upper midwestern United States. Short-distance migrants overwintering in

Globally, birds have advanced the start to their breeding seasons by forty to sixty days since the early nineteenth century.

southern states, such as pine warblers, sensed the early start to spring and initiated their migration northward earlier than they had in recent years. In contrast, long-distance migrants such as blackpoll warblers overwintering in South America waited for days to lengthen before initiating their journey. These birds had no idea what was happening in the Northern Hemisphere and arrived at their breeding grounds at the same time as usual. By the time the blackpoll warblers reached their breeding grounds, their primary food source—caterpillars—had already peaked. As a result, the birds' population numbers suffered.[20]

It gets even more complicated. In the Northern Hemisphere, the pace at which plants are undergoing leaf out across the country from south to north—referred to as the "green wave"—is also increasing.[21] What this means is that the number of days that pass between when plants leaf out in Georgia and when plants leaf out in Maine is shorter than it used to be. The implication is that all birds migrating north across the United States—regardless of whether they are migrating from the Southern Hemisphere or simply from the southern United States—must move more quickly to keep pace with food availability along the migration route and arrive at their breeding grounds when food is available. Some short-distance migrants such as eastern phoebes and American robins are responding to conditions along their migration route and speeding up their journeys. Species such as great crested flycatchers, indigo buntings, and scarlet tanagers, however, are not keeping up with changing conditions, thereby risking serious declines in their numbers.[22]

Successfully reaching summer grounds does not mean it is time to rest; after this milestone is reached, it is time to breed and incubate eggs. These activities are also occurring earlier in the spring than they used to. Since the late nineteenth century, egg-laying dates have advanced on average by ten days in the United States, though some species show much greater advancements.[23] In the upper midwestern United States, American kestrels now lay eggs more than fifty days earlier than they did a century and a half ago, and grasshopper sparrows have advanced their egg laying by forty-four days. Of seventy-two bird species evaluated in one study, only one—the American robin—demonstrated a significant delay in nesting time.

Many birds are also extending their breeding season later in the year—that is, remaining reproductively active for a longer period—than they have in the past. This is especially true for bird species like reed warblers and black-throated blue warblers that lay multiple clutches of eggs in a season.[24] More clutches per season is a good thing for these birds' population numbers because it means more baby birds. In contrast, birds that lay only a single brood in a season, such as sandwich tern, are experiencing an abbreviated breeding season, with a poorer outlook for survival.

Mammals—Both Large and Small—Are Also Shifting Their Active Periods

Birds are not the only group of animals showing clear changes in the timing of their activity. While mammals are generally trickier than birds to observe, evidence of shifts in the timing of their seasonal activity is also piling up. Because many small mammals like rodents and bats are secretive, scientists must get creative with the techniques they use to monitor these critters. One example of a clever approach originates from Texas, United States, where a group of researchers used weather radar to investigate patterns in the springtime emergence of Brazilian free-tailed bats from their overwintering cave. The radar data showed that between 1995 and 2017, the bats' springtime emergence advanced by approximately two weeks.[25] Similarly, in many parts of the world, mice, squirrels, marmots, hamsters, and bats are all emerging from winter hibernation earlier now than in the past because of warmer temperatures and earlier snowmelt. This ahead-of-schedule appearance can be a problem when the plants and insects they depend on for food aren't yet available for eating.

While large mammals like bears are bigger and easier to observe, tracking their phenology presents some of its own challenges. Bears are primarily tracked using tags or collars, and getting those devices onto the animals is neither inexpensive nor simple, so monitoring is more limited. Such efforts reveal that black and brown bears in North America and Russia are now emerging from winter hibernation earlier than in previous decades.[26] Zoos around the Northern Hemisphere similarly report bears emerging much earlier than normal in years with warm winters.[27]

Bears' earlier emergence from hibernation is driving an increase in conflicts with humans. Bears' food resources are not necessarily available when bears emerge earlier than usual, forcing them to supplement their diets. Often this involves breaking into dumps, houses, and cars to forage for snacks like cheese puffs and macaroni. Warmer temperatures are also pushing black and brown bears to delay their entry into hibernation in the fall. Longer periods of activity means that the chances for bears to enter into a conflict with humans is greater.

Earlier springs are affecting mammals' ability to successfully produce offspring too. Researchers studying reproduction in Richardson's squirrels in Manitoba, Canada, observed that more than half of the males possessed nonmotile sperm—nonswimmers—in spring 2012.[28] The reason behind this surprising phenomenon was the very warm conditions experienced across North America that year. Male squirrels typically emerge several weeks from hibernation earlier than females, and this extra time allows them to prepare developmentally for the mating season. In 2012, the emergence of male and female squirrels was separated by only a few days, and the males' bodies were not yet ready.

Cold-Blooded Creatures Show Especially Large Changes

Changing temperature conditions have especially strong impacts on cold-blooded animals like amphibians and reptiles that can't regulate their body temperature. Mild winters can trigger earlier emergence from brumation—hibernation in cold-blooded animals—and subsequent exposure to cold spells can damage their tissues or lead to death. Rising global temperatures are predicted to be especially hard on cold-blooded animals, increasing the risk of overheating as well as limiting periods suitable for finding food, mating, and engaging in other activities.

Do you remember watching tadpoles transform into frogs in an aquarium in your elementary school classroom and dissecting frogs in high school? If so, you might remember that amphibians like frogs are distinguished by a gill-breathing larval stage followed by a terrestrial, lung-breathing adult stage. These adaptations make them further vulnerable to increasing temperatures due to their permeable and exposed skin and shell-less eggs. On a global scale, amphibians have shown some of the strongest responses to increasing temperatures, having shifted their phenology by at least twice as much as other taxa, at a rate of approximately six days per decade. Though data are limited for many species of amphibians, a few findings really stand out. In South Carolina, United States, tiger salamanders and ornate chorus frogs now initiate breeding nearly sixty days—two months!—earlier than in the 1970s. Autumn-breeding dwarf salamanders have shifted their breeding by a whopping seventy-six days later than in the 1970s, mainly in response to changes in temperature.[29]

Reptiles, a group of entirely terrestrial animals encompassing turtles, crocodiles, lizards, snakes, and tuatara, are likewise exhibiting shifts in their periods of activity. Though studies of phenology are also limited, general patterns indicate earlier springtime activity in temperate environments with warmer temperatures. In the United Kingdom, the European adder has advanced springtime emergence by twenty-eight days since the 1980s. The earlier springtime emergence puts the snakes at greater risk of injury due to freezing temperatures.[30] Changing phenology is only one factor contributing to reptiles' and amphibians' declines; moreover, these animals face major challenges of habitat fragmentation, disease, harvesting, and competition from invasive species.

Amphibians have shown some of the strongest responses to increasing temperatures, having shifted their phenology by at least twice as much as other taxa.

Insects, also cold-blooded creatures, are similarly highly sensitive to the temperature of the surrounding environment. In years with warmer springs, many insects are active earlier in the year. In China, the United Kingdom, and the United States, many insects show earlier activity in response to warmer temperatures. Species that are active earliest in the season—typically those that overwinter as adults—show the strongest responses. For example, the green-veined white, a widespread butterfly in Europe, Asia, and North America that frequents meadows, is among the earliest butterflies to appear in the spring. In Sweden, the average flight time for these butterflies advanced by nearly a week over the period 1991–2010, though some species, such as the dingy skipper, shifted earlier by nearly four weeks.[31]

The insects showing the greatest advancements in spring activity are often the ones extending their season later in the autumn as well. In at least some cases, this is the result of additional generations getting squeezed into the growing season. The American grass-leaf aphid, an apple-green sap-sucking insect only a couple millimeters in length, has both advanced the timing of its first flights in the spring and delayed its last flights by several days since the 1960s in the United Kingdom. This advancement is a great deal more than other species of aphids under observation by the Rothamsted Insect Survey, a long-standing aphid-monitoring program in the United

Kingdom.[32] The secret to this insect's success? It is anholcyclic: there are no males. All individuals of this species are females that can lay eggs without the contribution of genetic information from males. Without the limitation of males, this species can take advantage of the longer growing season by producing more offspring, showing the greatest increase in population size among the fifty-five aphid species tracked through the survey.

In contrast, insects that initiate their activity later in the season show a weaker response to increasing temperatures, and in some cases are even delaying their activity. For example, two species of ground beetles, the violet ground beetle and black-headed beard runner, demonstrated a shift toward later springtime activity in Scotland over a twenty-five-year period.[33] At the same time, both species evidenced large drops in population numbers, suggesting that these insects are not well suited to the novel conditions emerging in the region. Similarly, the rock grayling butterfly, which first emerges in late July in Sweden, has significantly delayed its first flight time in recent decades.[34] One explanation might be that many insects require exposure to a certain amount of cold in the winter to fully enter winter diapause, a state of arrested growth that insects enter over the cold winter months. If a winter is particularly warm and the insects miss out on their necessary chill, they may actually require more springtime warmth to break out of diapause and thus ultimately delay their springtime emergence.

Aquatic Animals Are Responding to Changing Water Temperatures

As you might expect, aquatic critters aren't being spared from changes in their phenology either. Warmer global temperatures are leading to less winter ice cover on lakes and rivers, earlier springtime ice melt, and warmer water temperatures. When lakes are warmer, the amount of oxygen dissolved in the water decreases, nutrients like phosphorus and nitrogen increase, and lake turnover—when colder water at the bottom of the lake mixes with warmer water near the surface—occurs earlier in the year. In response, many freshwater organisms are shifting the timing of their seasonal activity. Freshwater invertebrates, such as dragonflies and damselflies, are advancing emergence from their aquatic, larval stage. Many freshwater fish are now spawning earlier as well. In Lake Michigan, the second largest of the five Great Lakes in the northern United States, yellow perch are spawning earlier in the spring and lake trout are spawning later into the fall.[35]

Sea creatures are similarly showing clear changes in their phenologies. As in other systems, the rates of change vary among taxa, though all through the food web, changes are afoot. Plankton—small and microscopic plants and animals that live their brief lives adrift in water—are highly sensitive to water temperature. Worldwide, spring plankton blooms—when the organisms grow rapidly and often become so abundant that the water changes color—have shifted anywhere from a week earlier than in the past to three months for the water flea, *Bythotrephes longimanus*, in Lake Maggiore, Italy.[36] Similarly, the time when baby fish hatch from their eggs is changing, though not consistently among fish species. About half of forty-five fish species evaluated in a study off the eastern coast of the northeastern United States demonstrated changes in fish egg hatch. For the most part, winter and spring season fish shifted egg hatch earlier in the year. Fish species that hatch their eggs in the summer and fall seasons generally shifted their activity later in the year.[37]

When aquatic creatures move around is also changing. An impressive set of observations collected on a weekly basis in Narragansett Bay, Rhode Island, over a span of seven decades without interruption shows clear changes in the arrival and departure times of migratory fish and squid. Since the late 1950s, the longhorn sculpin, a small, ray-finned fish that spends winters in the bay, has shortened the length of its wintertime stay by a whopping 118 days.[38] Winter resident fish including ocean pout and red hake have similarly shortened the duration of their visits by over 90 days. Nearly all summer resident fish monitored now, including scup, butterfish, summer flounder, and striped searobin, remain in the bay longer in the summer by 30 to more than 80 days. As well, the longfin squid has extended the length of its summertime stay by nearly 85 days.

Similar changes have been observed for river fish. Anadromous fish uniquely spend most of their adult lives at sea, but migrate up freshwater rivers to spawn. Migrations in Atlantic salmon have shifted two to three weeks earlier than in the 1970s in the North Atlantic in response to warmer stream temperatures. Alewife, another anadromous fish common in North America, has likewise shifted their migration noticeably—by over three weeks in Maine over just two decades.

Shifts in the timing of activity among marine mammals are beginning to emerge too, especially among those attuned to autumn sea ice "freeze-up." The timing of sea ice freeze-up in the Arctic has shifted later at a remarkably rapid pace in recent decades; in the Chukchi Sea of the Pacific Ocean, dates have shifted at least one week later per decade since the 1970s.[39] Nearly in lockstep with this change, beluga whales have delayed the initiation of their autumn migrations from their summer foraging habitat in the Chukchi Sea to their Bering Sea winter habitat by two to four weeks. In the nearby Beaufort Sea, however, belugas have not changed the timing of their migration, despite clear delays in local ice freeze-up. Narwhals—toothed whales recognized by the large tusks protruding from their heads—delayed their autumn migrations from nearshore ice-free locales to ice-covered wintering grounds by twenty-one days in two decades.[40]

Impacts of warmer water temperatures extend to aquatic insects. Aquatic insects spend some portion of their life cycles—typically larval stages—underwater. During their life stages spent underwater, they are often highly sensitive to water temperatures. In general, many aquatic insects are developing more rapidly and emerging from the water earlier in the season under warmer conditions. For example, in Oregon, United States, caddis flies, mayflies, and stone flies all emerged when the air and temperatures were warmer. Yet crane flies—also known as "mosquito hawks" for their long, spindly legs and a body shape that resembles a mosquito—are responding in surprising ways to warmer air and water temperatures in Germany. Warmer water temperatures slow the development of their larvae, shifting emergence later in the spring.[41] Furthermore, warmer temperatures are prompting the insects, which neither eat mosquitoes nor bite humans, to grow longer wings and heavier bodies.

Species That Are Shifting Their Timing Are Winning

Adapt or die. It's a common aphorism repeated in the business world, with origins in biology, referring to the eventual demise of species that aren't able to keep pace with changing conditions. This truism extends to changes in phenology as well. In general, species that are changing the timing of their activity in response to rapidly changing environmental conditions—for instance, advancing their springtime activity to keep up with increasing temperatures—are faring better and thus becoming more abundant on the landscape. Conversely, an inability to undergo leaf out or flowering earlier in the season in response to increasing temperatures generally spells danger for plants. Records collected by Henry David Thoreau in the 1850s and recently resampled by researchers at Boston University have revealed that species such as Saint-John's wort and Indian pipes, which have not advanced the timing of their springtime activity since Thoreau observed them, are becoming far less common in the landscape, while yellow sweet clover, Pennsylvania smartweed, and black swallowwort, which have advanced their flowering times steadily with increasing temperatures, are not decreasing in abundance.[42] Likewise, Fender's blue butterfly, an endangered butterfly with a distribution restricted to five counties in Oregon, has consistently advanced the timing of peak flight in recent decades. In several locations, butterfly numbers are increasing—a good sign for this species' persistence.[43] In Finland, species that have both shifted their ranges northward and advanced their flight season in response to warmer temperatures have showed population increases, whereas species that have not changed either their range or timing of activity have shown clear declines in numbers.[44]

Are There Other Reasons We Could Be Seeing Changes in Phenology?

Temperature far and away exerts the most influence on the timing of when plants and animals undergo seasonal changes. Because so many seasonal events in plants and animals are tightly tied to temperature, we have witnessed abundant evidence of changes in phenology as global temperatures have increased. There are reasons other than temperature increase, however, that might lead to changes in phenology. As we explore further in chapter 5, environmental variables like daylength, available moisture, and water temperature also frequently play important roles. Moreover, the timing of activity is shaped by what other species are doing. And finally, other forms of global change such as urbanization and habitat fragmentation shape phenology too. Urban areas, with high degrees of heat-emitting asphalt and dark surfaces, are warmer than surrounding, undeveloped landscapes. As a result, plants in urban areas leaf out earlier in the spring and delay leaf color change later into the fall. The abundant artificial light in cities also affects the timing of activity in plants and animals by tricking them into believing the days are longer than they are, leading them to undergo spring activity earlier and extend autumn activity later in the year.[45]

You may have heard the weather forecaster on the local news broadcast talk about this being an "El Niño year." What does that mean? The earth's climate undergoes regional oscillations that last from a few months to several decades, and impact local, regional, and global weather patterns. The oscillation we hear about most frequently is El Niño–Southern Oscillation, which is characterized by warmer sea surface temperatures in the equatorial Pacific Ocean. An El Niño year brings more rainfall to the southern United States, warmer conditions in the northern United States, and drier conditions in northern Australia and Southeast Asia. When the opposite set of conditions occur, with cooler than average temperatures in the equatorial Pacific Ocean, the event is known as La Niña. A La Niña year brings the opposite effects, including cooler temperatures in the northwestern United States, warmer winter temperatures in the southwestern United States, dry conditions in South America and East Africa, and increased rainfall in northern Australia and Southeast Asia. The changes in weather conditions that these oscillations bring can influence the timing of seasonal events. For example, the increased rainfall brought to Australia by La Niña events leads to longer egg-laying periods in many bird species.[46] Similarly, plants appear to green up earlier in the spring in El Niño years and later in La Niña years in the United States.[47]

The span of time evaluated in a study matters as well. The observational datasets scientists use to estimate changes in phenology are typically rather short in duration. While some records span up to 200 years, many are decades in length or even shorter. Datasets collected over a short period of time can lead to different conclusions about whether phenology is changing or not. For instance, an examination of the first flower dates for seventy-five low-elevation plants in southern Arizona, United States, collected from 1984 to 2003, suggested that only 12 percent of the species were shifting their flowering time, and more than half of these were delaying spring flowering during the period.[48] Yet a subsequent analysis of the same dataset that included an additional 13 years of observations (1984–2016) revealed that on average, plants advanced their flowering time by 2.5 days per year.[49] The different findings may reflect changes in the climate over the duration of the study. The 1980s were comparatively wet, while the early 2000s were much drier. Concurrently, temperatures steadily increased. Plants in this region are clearly responding rapidly to changing conditions. The important take-home message is that the duration of a phenology dataset as well as the years when the observations were collected both shape the findings.

Are There Limits to How Much Phenology Can Change?

The ability of organisms to adjust the timing of their activity from one year to the next in response to varying conditions is termed "plasticity." Largely, the changes in phenology we have observed, such as when a tree flowers earlier in a warm year and later in a cool year, or that migratory birds' arrival times vary from one year to the next, are the result of plasticity. Plasticity enables species to respond to varying conditions from year to year. Long-term phenology observations have revealed that many species are able to "track" changes in their environment by leafing out or flowering earlier in warmer years and later in cooler years. The species that have shown the largest changes in the timing of their activity, like blueberries in Massachusetts and tiger salamanders in South Carolina, are exhibiting plasticity in their phenology.

Because species are sensitive to different sets of environmental conditions, or cues, to trigger their activity, their ability to adjust the timing of their activity—that is, to exhibit plasticity—varies. Species that initiate growth in the spring simply in response to warmth, like lilacs and birches, are showing large changes in the timing of their activity. In contrast, beech and linden trees, both slow-growing tree species that populate mature forests, seem to require chill and daylength conditions to trigger springtime growth. Consequently, these trees have advanced their springtime activity to a lesser degree than other forest plants. Plants with a daylength requirement are even more constrained in how much they can advance their springtime activity because leaf out in these species just won't happen until the days reach the required number of daylight hours. Long-term phenology studies taking place in montane and arctic environments where conditions are changing most rapidly and are most extreme show signs of plants reaching their capacity for change, suggesting there might be limits to how much species can change their phenologies.[50]

Species can also adapt to changing conditions. This occurs when individuals that possess traits that are better matched to their environment—such as the ability to fly more rapidly to keep up with advancing food resources or leaf out in response to less warmth in the spring—pass their genetic material to subsequent generations. Because this process involves multiple generations, this takes time. There are a few cases where rapid adaptation appears to be occurring in response to rapidly changing environmental conditions. Annual plants, which complete their life cycles within a matter of months, are likely candidates for exhibiting rapid adaptation. Field mustard is an annual plant common to roadsides, ditches, and cultivated fields around much of the globe. Researchers experimentally limited plants' access to moisture during their growing season. In response, the plants' offspring consistently flowered earlier in the season. Such rapid evolution has been documented in insects too, another group of organisms with short generation time, though the examples are few. Still, the rate of global temperature increase is estimated to be approximately a hundred times greater at present than prior to the onset of the Industrial Revolution, when atmospheric carbon dioxide concentrations started to dramatically increase.[51] In many cases, species are unlikely to possess the ability to undergo multiple generations sufficiently quickly to stay within the bounds of novel climate conditions.

3

WHAT DATA AND INFORMATION DO WE USE TO EVALUATE CHANGES IN PHENOLOGY?

How can we evaluate whether the timing of activity in plants and animals has changed? Quite simply, we need to know when things happened in the past and when they are happening now. By comparing these reports, we can ascertain whether things have changed and by how much. Analysts have cobbled together records of when seasonal events occurred in the past from a vast array of sources to tell a more comprehensive story of change, including records collected through formal monitoring efforts, naturalists' diaries and notebooks, news and social media accounts, repeat photos, and satellite imagery.

Independent Efforts to Document Phenology

Direct reports of the timing of seasonal events in years past are among the most robust sources of information for determining whether the timing of seasonal events has changed. Some of the most valuable records of this sort were collected by individuals simply because they liked to record the changing of the seasons. For example, Ms. Jean Combes has documented the timing of leaf out in four tree species—oak, ash, horse chestnut, and common lime—since 1947 merely because she "loved nature," yielding an uninterrupted record spanning nearly seventy-five years.[1] Such long-term records are invaluable for documenting whether things like the timing of leaf out have changed. Combes's observations show that oak bud burst has advanced by nearly a month under her watch.

In the US state of Arizona, Dave Bertelsen has hiked the ten-mile out-and-back Finger Rock trail in the Santa Catalina Mountains north of Tucson multiple times a month since 1984, with few breaks. On every hike, Dave has documented every species of plant he observes in flower within each of the five miles of the trail, resulting in nearly two hundred thousand observations collected during more than eighteen hundred trips.[2] Though Dave originally began documenting what he saw on his hikes because he noticed different plant assemblages at different elevations, he persisted in the task because he started to note changes. His observations have proven incredibly valuable, serving as one of the richest long-term records of flowering from a semiarid environment anywhere in the world, and are especially useful because they capture change occurring along an elevation gradient of more than four thousand feet (twelve hundred meters). Dave's original intuitions were correct: things are indeed changing in this environment. His observations showed that Parry's false prairie clover, a shrubby plant with small purple flowers, and sixweeks threeawn, an annual grass, both advanced flowering time by more than a month over a twenty-year period.[3]

One of the most well-known examples of historical phenology records hails from England, where Robert Marsham, an English naturalist, documented leaf out, flowering, and springtime arrival in dozens of plants and migratory birds starting in 1736. These observations were continued by five generations of the Marsham family, resulting in a phenological record spanning over two hundred years (1736–1947) with few gaps. The Marsham family's observations have revealed that springtime events like flowering in several species of plants along with the first spring appearances of birds, butterflies, and frogs are cued by spring temperatures. Robert's extensive, colorful accounts also included a mention that his chamber pot regularly froze overnight during the very cold winter of 1739. The Marsham family's records have served as a crucial baseline for comparing contemporary phenology observations and identifying changes.

In the United States, another well-known historical figure's detailed observations—similarly collected out of personal curiosity—are a rich and invaluable resource for evaluating how things have changed. Thoreau, best known for *Civil Disobedience*, his involvement in the Underground Railroad, and his account of living in a ten-by-fifteen-foot cabin on Walden Pond for two years, meticulously documented flowering, leaf out, fruiting, and bird arrival in dozens of species in the mid-1800s. In 2003, Boston University professor Richard Primack discovered these records through conversations with colleagues and historians. Dr. Primack and his graduate students have resampled Thoreau's haunts every year since 2003. They combined Thoreau's historical records, their own contemporary observations, and reports of phenology collected by others in the intervening years to tell a compelling story of change. Trees in the area generally leaf out two weeks earlier now than they did in Thoreau's time, and wildflowers bloom approximately one week earlier. Not all species are advancing, however; while warbling vireos arrive at Walden over a week earlier than in Thoreau's time, barn swallows and ovenbirds now appear over a week later.

Records of phenology captured in diaries or notebooks continue to surface, providing invaluable windows into the past. The diary of a dedicated naturalist in southeast Massachusetts, Ms. Kathleen Anderson, revealed that over the period 1970–2022, nearly all the spring events she observed, including the arrival of wood ducks, house wrens, and chipping sparrows, the first *peep* of spring peepers,

Records of phenology captured in diaries or notebooks provide invaluable windows into the past.

and first flight in mourning cloak and spring azure butterflies, shifted earlier.[4] The records of leaf out, flowering, and migratory bird arrival kept by L. S. Quackenbush, a hunting guide in Aroostook County, Maine, in the mid-twentieth century, show that spring arrival in the white-crowned sparrow and least flycatcher have shifted earlier by nine days as temperatures in the region have warmed.[5] In Ohio, researchers recently resampled records collected by Thomas Mikesell, a farmer who carefully documented the timing of leaf, flower, and fruit activity in dozens of fruit trees, crop plants, and native trees, shrubs, vines, and flowering plants from 1883 to 1912. Since Mikesell's collected his first observations 130 years ago, autumn leaf color change in elm, black walnut, white oak, and sumac trees has drifted nearly two weeks later.[6]

Historical and Contemporary Phenology Monitoring Efforts

In the seventeenth, eighteenth, and nineteenth centuries, plant phenology was formally tracked in several countries to support farming practices, spurred on by the growing appreciation of the relationships between temperature and other environmental factors and plant growth. These records are an invaluable resource for determining whether the timing of seasonal events has changed. In 1760, the formal observation of plants and meteorologic parameters was initiated in Bern, Switzerland, in what is believed to be the earliest such effort. In the United States, several regional efforts launched in the mid-twentieth century centered on tracking the phenology of cloned lilacs and honeysuckles. The effort spread into Canada, and for several years in the 1970s, observations were collected at over three hundred sites. In China, monitoring was initiated in 1934, and managed by the Chinese Academy of Sciences off and on for several decades. In 1980, the Chinese Meteorological Administration established an additional network with an agricultural focus. Organized phenology monitoring in Japan began in the mid-1950s, with the aim of monitoring the local climate through seasonal phenomena.

The formal tracking of plant phenology was also initiated in various locations in support of another popular activity of this age: naming, comparing, and classifying. The Royal Meteorological Society of London started monitoring plants, animals, and insects in 1857; this monitoring was sustained until the mid-1900s. The Royal Society of Canada similarly launched an effort to track 170 events including plant bloom times, bird arrivals, and ice formation and breakup in 1892, with the stated aims of "collation and publication." In the United States, a network of academic institutions in the state of New York collected phenology and weather observations at over seventy sites from 1826 to 1872. As well, two formal programs—the

Chronicles of Nature (Letopisi Prirody) monitoring program, in which paid professionals collected observations, and Fenologicheskii Klub, in which volunteers submitted observations, have tracked phenology in plants, birds, insects, amphibians, reptiles, and fungi in the Russian Federation, Ukraine, Uzbekistan, Belarus, and Kyrgyzstan with few interruptions since 1890. Historical and long-running monitoring schemes such as these provide invaluable information that is foundational in establishing how things have changed in recent decades.

Two networks established in the late 1950s in Europe and the United States coordinated observations of cloned plants. Observations of cloned plants, which are genetically identical, are especially useful because any variation in when individual plants undergo leaf out or bloom can be attributed to differences in environmental conditions such as temperature rather than to genetic differences that might affect phenology. The International Phenological Gardens network, which has been in operation since 1957, coordinates observations of dozens of cloned plants across Europe. In the United States, multiple regional efforts engaged volunteers in tracking leafing and flowering in three cloned plants: the Red Rothomagensis lilac and two varieties of honeysuckle starting in the 1950s. Like many other monitoring efforts, these networks were originally begun with the aim of improving crop yield. In recent years, the observations amassed through these networks have supplied some of the most widely referenced reports of changes in the timing of leaf out and flowering.[7]

As an appreciation of phenology as an indicator of how organisms are responding to rapid change has grown in recent decades, phenology monitoring networks have spread. The Global Phenological Monitoring Programme was initiated in the early 1990s in several European countries to better track plant response to climate change. Observations are collected on fruit trees and ornamental shrubs in designated gardens by professionals, students, and volunteer observers. The PhenObs Network, which focuses on tracking phenological events in herbaceous (nonwoody) plants—typically underrepresented in phenology monitoring efforts—was kicked off in 2017 and similarly motivated by changing climate. This network tracks plant phenology at botanical gardens, as these sites are permanent, easily accessible, and maintain plants in optimal growing conditions. Finally, in many countries, "citizen science" phenology monitoring programs have been created to document the response to rapid global change. In these efforts, phenology data collection is primarily undertaken by volunteers and students. Volunteer-oriented phenology monitoring programs are currently active in many countries worldwide, including the United States, United Kingdom, Canada, Australia, Bhutan, India, Sweden, Netherlands, Japan, and Ireland; these programs are described in further detail in chapter 6.

Many monitoring programs track various aspects of animal populations, such as presence (where they are) and abundance (how many of them are there), and trends in population size are another potential gold mine of phenology data. Though not established with the explicit focus on questions pertaining to phenology, in many cases, the data gathered through these efforts have provided critical pieces to the puzzle of understanding how phenology is changing and why. In the United Kingdom, the Butterfly Monitoring Scheme, established in the 1970s to monitor the abundance of butterflies, has yielded data invaluable for assessing changes in first spring appearances. When combined with historical butterfly observations extending a hundred years back in time, these observations reveal advancements of two to five weeks in first appearances from the 1970s to the turn of the twenty-first century. The Butterfly Monitoring Scheme has since been extended to nearly two dozen countries in Europe (butterfly-monitoring.net), and observations reveal clear shifts in flight times in these other countries as well.

Satellite and Aircraft-Borne Sensors: Enabling a Bird's-Eye View of Phenological Change

The advent of satellite-based sensors in the 1970s brought a whole new dimension to tracking phenology, enabling

As an appreciation of phenology as an indicator of how organisms are responding to rapid change has grown in recent decades, phenology monitoring networks have spread.

regular, repeated data collection in previously undersampled regions. Remote sensing refers to using sensors to collect data and information from afar. Sensors borne on satellites, aircraft, and unpiloted aerial vehicles record energy reflected by the earth's surface within various segments of the electromagnetic spectrum. The relative strength of these signals can be compared to estimate the "greenness" of the landscape. This approach most commonly involves comparing the amount of reflectance recorded in the red region of the spectrum, which green vegetation strongly absorbs, and the near-infrared region, which green vegetation strongly reflects. These measures form the basis of several commonly used vegetation indexes including the normalized difference vegetation index and enhanced vegetation index.

Measures of reflectance are collected within pixels of an image or scene. Depending on the sensor, pixels can encompass regions smaller than a square meter on the ground to multiple square kilometers. As such, many individual plants are lumped together into a single pixel. Further, because sensors are borne primarily on satellites orbiting the earth, data are not collected continuously. The return interval—the amount of time that lapses until the satellite repeats coverage of a location—varies substantially by platform, ranging from every day to once every several weeks.

By calculating a measure of "greenness" for a location over the course of the season, analysts can estimate the start and end of the growing season there. Many studies using remotely sensed information have revealed clear trends in the start of spring—dubbed the "green wave" in reference to the spread of green from the equator toward the polar regions in the spring season—and end of autumn, captured as the "brown wave." Remote sensing approaches also show promise for capturing flower events, phytoplankton blooms, and other seasonal events that cause the surface of the ground to change color.

Repeat Photography: Local-Scale Remote Sensing

In summer 2006, I was routinely awakened by the subtle *click-click* of a digital single-lens reflex camera shutter. My husband and I had established a setup in our bedroom to collect repeat photos, with the camera view fixed on a plot in the backyard where we had planted wildflower seeds. Back in 2006, inexpensive weather-ready cameras that could be programmed to collect photos at a specified interval were not readily available. My crafty husband devised a clever alternative: to control the camera with a laptop, and mount the camera inside and point it out the window. Over the ensuing months, the monsoon rains triggered the germination of grasses and wildflowers, and our camera captured photos of the plot on an hourly basis. We used these photographs, collected from a fixed position, to document the rate of growth in the plants as represented by "greenness," following an approach similar to that used to estimate greenness from satellite-borne sensors. Digital cameras collect information in three separate channels: red, green, and blue. The strength of the reflectance in these channels can be compared to estimate green plant biomass. Our study showed that regularly collected images of plants readily capture plant growth and key phenological transitions.[8]

Our setup, consisting of a laptop running continuously indoors controlling a digital single-lens reflex camera aimed out a window, is clearly not practical to implement for any reason other than a demonstration. Recent technological innovations have dramatically increased the options for programmable cameras, including inexpensive game cams, trail cams, and bird cams (e.g., wingscapes.com and moultriefeeders.com) that can now be readily deployed in remote areas, powered by battery or solar panels and can store images on memory cards for months at a time. Images sourced from these inexpensive cameras are increasingly used to determine the conditions that cue green up, flowering, and leaf color change in various plant species in a range of environments.

In a similar fashion, there is an extensive network of fixed cameras collecting digital images at the landscape level around the world as part of the PhenoCam Network. Over four hundred sites collect digital photos on an hourly basis and contribute them to an archive that houses over seventy million images. The PhenoCam Network started with a handful of webcams mounted on flux towers—towers instrumented with many sensors measuring carbon dioxide, water, and energy in the atmosphere—because the researchers conducting this work wanted to have nice pictures of the landscape to include in their presentations, and know when leaves appeared in the spring and dropped in the fall.[9] The practice continued—and expanded wildly—because the cameras not only yielded pretty pictures of the trees throughout the growing season but also a unique and valuable data stream. The digital pictures could be examined to show green up, brown down, impacts from insects or disturbances like major wind events, and subsequent recovery from these events. The European Phenology Network and the Phenological Eyes Network in Japan similarly collect imagery at a network of sites using standardized equipment and protocols.

Creative Ways of Tracking Animal Phenology

My friend Dr. Katy Prudic, an entomologist and professor at the University of Arizona, studies butterflies and how they are being affected by climate change. She and her butterfly chasing colleagues have determined that many butterflies in the western United States are both emerging earlier in the spring and declining in numbers.[10] How do they know these things? They make frequent visits throughout the spring to popular butterfly gathering spots and sweep for the critters using big nets. They then compare what they find to when the same insects were seen in flight in springs of the past—collected by both professionals and amateur naturalists. Determining whether the timing of activity in animals has changed requires historical and contemporary observations, and as with plants, observations collected by professionally trained scientists and hobbyists using a range of techniques serve as crucial pieces of this puzzle.

The Monarch Watch program, based in Kansas (monarchwatch.org), employs a simple approach to track monarchs on their migration across North America: stickers. Circular, adhesive tags bearing unique codes are gently affixed to the wings of individual butterflies that are netted at various locations on their migration south in August, September, October, and November to their overwintering locations in central Mexico. When the butterflies are recaptured, their tags are logged into a central database, along with the location and date of the sighting. Established in 1992, this effort has generated tens of thousands of observations of monarchs' whereabouts and provides critical information supporting the management of this imperiled species.

Migrating butterflies can also be tracked using tiny transmitters—weighing less than a quarter of a gram—glued to their abdomens. Data from these units have helped scientists track the North American fall monarch butterfly migration, revealing that the butterflies travel an average of sixty-one kilometers a day.[11] Recent advances in technology have dramatically improved our ability to track precise locations and movements of terrestrial and aquatic animals. Birds, small mammals, snakes, tortoises, and bats can all be tracked using telemetry—mounting transmitting devices on animals and tracking their movement. Telemetry units are often glued onto the head or backs of snakes, tortoises, and organisms with tougher skin, whereas bats and birds can be outfitted with tiny "backpacks" bearing the equipment. Larger mammals frequently wear their sensors as collar, ear, or tail tags. Tiny devices can be implanted under the skin too. The radio signals transmitted by these devices can be tracked in the field by a technician using a handheld receiver as well as by drones and satellites.

Just a short time ago, an antenna was mounted on the International Space Station that now tracks the location of more than eight hundred species of animals, from bats to elephants. Animals are outfitted with special tags that transmit a signal that the International Space Station can receive. In 2020, data collected by this system revealed that rather than flying south to their wintering grounds in the Antarctic after nesting, arctic terns moved north of their breeding site for a few weeks. Avian ecologists proposed that this surprising move by the terns allowed them to fatten up prior to their long journey south.[12]

The sounds that animals like birds, bats, marine mammals, amphibians, insects, and even whales and dolphins make to communicate with each other as well as navigate indicate when they are active. Male frogs vocalize to draw in a mate; bats echolocate—navigate by the echo return of sounds they utter—to find food in the dark. These sounds are another rich source of information about where and when various species are present and can be tracked using acoustic monitoring techniques. Using bioacoustics, Australian researchers demonstrated that both insects and birds vocalize less frequently when temperatures are warmer than usual, suggesting that as temperatures warm, these species could face more challenges finding mates and reproducing.[13]

Bird bands are another rich source of information. Birds are commonly tracked using rings or bands affixed around their legs. The bands are engraved with unique sets of numbers, which are housed in central databases. While birds were tracked in various ways as long ago as 200 BC, the practice of outfitting birds with their own version of license plates as we know it today was initiated at the beginning of the twentieth century. Bird banding data, presently overseen by federal governments and consortia around the world, have enabled a much better understanding of bird migration patterns and how they are changing. As just one example, a US-based research team evaluated nearly three million banding records of wood warblers over the period 1961–2018 to evaluate the impact of temperatures on spring and fall migration dates. Sixteen of the nineteen species were documented earlier in the year when temperatures were warmer, whereas fall migration was sped up for only three of the birds in warmer years.[14]

Using Weather Radar to Track Animal Movements

Weather radar is used by meteorologists to detect rain, hail, snow, and sleet. Yet because the sensors work by sending signals and then listening for them to return when they bounce off solid objects in their path, they also detect airborne living things. Data collected by the NEXRAD radar network in the United States, consisting of 160 units distributed across the country, have been used to document seasonal emergence and flight in birds, bats, and aquatic insects. For instance, mayflies are insects that spend most of their lives underwater, emerging as adults only to mate and then die. Mayflies undergo synchronous emergence, meaning that nearly all individuals in the population transition to their flying adult stage at the same time. This event is cued by water temperature. Once streams warm beyond a critical threshold, nearly twenty billion insects can hatch in a single night along the Upper Mississippi River. Using radar data, scientists have determined that the number of insects has been declining sharply in recent years, with clear implications for fish and other species dependent on them as food.

In another creative application of weather radar data, a team of researchers from across the United States showed that the timing of the peak migration of insect-eating birds has shifted earlier in both spring and autumn in North America over a recent twenty-four-year period.[15] The enhanced understanding of how birds move through the landscape offered by the radar data is providing valuable insights into conservation strategies, revealing the importance of tracts of deciduous forests as key sources of food for these birds.

Inspired by the success of weather radar to document animal presence, a group of researchers in the United Kingdom constructed "vertical-looking" radar sensors with the specific aim of documenting insect size and movement. The insect-specific sensors are able to document smaller insects that weather radar do not detect. This system, called the BioDAR, has revealed that 3.5 trillion insects inhabit the skies of southern England, and lays a foundation for documenting changes or long-term trends in the timing or abundance of insects in the sky.[16]

Biological Collections

Collecting and preserving plant specimens has been practiced worldwide for hundreds of years, initiated as a matter of convenience. Back in the 1700s, when explorers' ship-based voyages stretched for months or years, space was at a premium. Pressed plants, collected to document and study the plants of a region, could be preserved for long periods of time and transported many thousands of miles by boat. Explorers such as Charles Darwin and Alexander von Humboldt collected tens of thousands of specimens that are still in existence today. These records are foundational in documenting plants' geographic distributions, understanding plant structures, and establishing relationships among various species.

Pressed plant specimens, housed in herbaria—collections of dried plants—are typically labeled with key information: the species, location and date of collection, and notes on the habitat where it was collected. Specimens are often collected when plants are in flower or bearing fruit, as these structures assist with plant identification. As such, each pressed plant specimen—or herbarium record—serves as a report of where and when a particular plant was present as well as information regarding its phenological status at the time of collection.

Though herbarium records were not originally collected with the express purpose of documenting organism's developmental status at the time of collection, they are a highly valuable resource for determining how plant phenology has changed. Hundreds of millions of herbarium records exist worldwide, and many specimens were collected prior to the Industrial Revolution and the consequent changes to global temperatures. As a result, they are well suited for documenting plant response to changing climate conditions, and recent analyses using herbarium records have revealed trends toward earlier flowering in locales around the globe. In addition, herbaria frequently house records from regions underrepresented in long-term monitoring, including the tropics and Southern Hemisphere.

A great deal of effort is being made to increase the accessibility of herbarium specimens for research. The first step is to digitize the specimens, which involves capturing digital images of each specimen and its associated information, including the date and location of collection. The growing resource of digital herbarium specimens has supported an improved understanding of plant responses to global change, including the spread of invasive species, shifts in population numbers in rare plants, changes in species geographic distributions, and the impact of various environmental pollutants. Capturing the phenological status of the specimens, meaning the presence of leaves, flowers, and fruits—foundational for phenology research—takes herbarium specimen annotation a step

Though herbarium records were not originally collected with the express purpose of documenting organisms' developmental status, they are a highly valuable resource for determining how plant phenology has changed.

further. This involves documenting the phenological status of the specimens—typically, the presence of leaves, flowers, or fruits—as well as in some cases the abundance of these structures—for example, the count of fruits present on the specimen. To date, most of this work has been performed manually by technicians and by volunteers in the CrowdCurio and Notes from Nature citizen science projects. Machine learning methods, which entail training a computer to make judgments about the presence and abundance of leaf, flower, and fruit structures, are being implemented increasingly widely to reduce the burden, dramatically speed up the process of digitization, and enable more research.

Natural history collections maintained by natural history museums and academic institutions house not only pressed plant specimens but animal specimens as well, including pinned insects, eggshells, bird nests, and taxidermy birds, reptiles, amphibians, and mammals. An estimated 1.1 billion specimens are housed in such facilities worldwide—a veritable Pandora's box to be discovered—though the majority still await digitization. Researchers are beginning to dip into the rapidly growing pool of digital specimen records, answering previously unaddressed questions about species' phenologies and how things are changing. Drs. Bryan McLean and Robert Guralnick, for instance, reconstructed breeding time in North American deer mice across the continent by combining digitized museum specimens with field censuses.[17] Using these data, the researchers determined that both temperature and daylength cue breeding in these critters, and as a consequence, the timing of breeding is starting to change. Similarly, a New Jersey–based research team led by Dr. Nicholas Dorian used digitized museum collections of 70 species of solitary bees collected from the northeastern United States and southeastern Canada spanning 168 years. Solitary bees are not part of a hive; they live alone, collect their own food, and lay their own eggs. The specimens revealed that since the mid-1800s, the bees have advanced the start time of their spring activity by approximately a week, and since the 1970s, the rate of change has increased fourfold.[18] The bees have also extended their period of activity by about a week, providing pollination services for a longer period than in previous centuries.

Newspaper and Social Media Accounts

Today, my Twitter feed informed me that wisteria is flowering near Perth and sea eagle eggs are on the verge of hatching at the Sydney Olympic Park, also in Australia, catalpa trees are in bloom in Nova Scotia, and swallows are nesting in the United Kingdom. Social media reports such as these as well as historical accounts of events captured in the traditional news media are valuable sources of information for increasing our understanding of phenology along with how it is changing. South African professor and researcher Dr. Jennifer Fitchett combs through historical newspaper articles and social media posts on Instagram, Flickr, and Twitter to reconstruct a timeline of when charismatic seasonal phenomena occurred in years past. Fitchett has used this approach to quantify shifts in the springtime bloom of the Namaqualand daisies in western South Africa, characterized by unbroken carpets of orange on an otherwise barren landscape; the midsummer migration of the brown-veined white butterfly, when tens of thousands of the small butterflies pass through Johannesburg on their way to Mozambique; the wildly purple spring bloom of jacaranda trees in South Africa's Gauteng Province; and the winter sardine run, when large schools of the fish migrate close to the KwaZulu-Natal coastline along South Africa's east coast.

Using a media record encompassing more than a hundred years, Fitchett and her team documented that the brown-veined white butterflies' migration through Johannesburg now occurs approximately thirty days earlier than in the early 1900s.[19] The same approach revealed a seven-week shift in the timing of jacaranda bloom, from mid-November in the 1920s and 1930s to early September in recent years.[20] Media reports also reveal that the Namaqualand daisies now bloom approximately twenty days earlier than in the mid-twentieth century, and the

Social media and historical accounts captured in the news media are valuable sources of information for understanding how phenology is changing.

sardines now arrive more than a week later than in the 1940s.[21]

Other Unexpected Sources of Information

Historical photos can be a helpful piece of the puzzle too, offering a literal snapshot of what plants were doing on the date the photo was taken. When photos include permanent recognizable features such as buildings and monuments, they can be revisited, and plant development status in the past can be compared to current patterns in the same species. Likewise, photos taken on the same day of year can reveal change. Photos collected at a World War I monument in London, England, on Armistice Day reveal clear changes over the past hundred years. Photos from the 1920s show bare trees, while recent photos are filled with green vegetation.[22]

Video footage collected at recurring cultural or social events can similarly be analyzed for clues regarding plant developmental status in years past. A group of Belgian researchers examined footage of the Tour of Flanders, a famous professional road cycling race in Belgium over a period of forty-five years (1981–2016).[23] They identified distinct individual trees along the racecourse, and documented the degree of development in leaves and flowers. In the 1980s, the trees were nearly always bare. By the time of the race in the latest years of the study, many of the trees were fully leafed out.

And what about all the surveillance cameras deployed across the landscape these days, including traffic cams, security cameras, and even doorbell cams? The rapid uptick in surveillance cameras offers a massive resource that for the most part has remained untapped for tracking phenology. A recent analysis of imagery recorded by traffic cams in the United Kingdom hints at the opportunity, demonstrating how techniques commonly used to determine the start and end of the growing season based on the image "greenness" commonly used with images from PhenoCams and other landscape-scale cameras can be applied to these novel image archives. Home security cameras, doorbell cams, and even automobile dash cams log imagery at regular intervals, frequently capturing the same field of view or organisms, and have potential for documenting phenology in vegetation.

Traditional Ways of Knowing Offer a Wealth of Valuable Insight

Indigenous peoples worldwide maintain invaluable information regarding how species and ecosystems are responding to changing environmental conditions, and increasingly, anthropologists trained in Western science practices are seeking reports of change from these communities. Because of their close ties to the land and ecosystem for survival, Indigenous peoples uniquely experience and understand changes. In a recent collaboration between Western-trained scientists and Indigenous communities to document evidence of recent change, elders representing Anishinaabek communities in the Great Lakes region of the United States and Canada reported that in their lifetimes, they have observed many shifts, including strawberries appearing in July now rather than June, a shorter fish-spawning season, and changes in the timing of bird migration, maple tree growth, and snake hibernation.[24] Similar efforts to record changes being observed by local communities are taking place around the globe.

Indigenous knowledge can fill critical gaps in locations where formal monitoring doesn't exist. For example, Indigenous elders from the Mbire District in Zimbabwe, where weather stations are sparse, reported changes in temperatures and rainfall patterns as well as delays in the flowering of trees such as Indian jujube and tamarind during their lifetimes, corroborating findings from long-term weather station data while adding nuance and specificity.[25] As several scholars point out, Indigenous communities offer not only information that complements traditional Western approaches to tracking change; local people's lived experience and intimate relationship with the land and other organisms in the area have the potential to identify changes that might otherwise go undetected.[26] Given this, efforts are underway to document such observations in ways that they can be incorporated into syntheses of climate change impacts like the periodic IPCC reports.[27]

The records of when seasonal events in plants and animals occurred historically, acquired from these diverse sources, tell a compelling story of how phenology is changing. These data also clearly show that changes are not consistent among species, locations, or seasons. Many gaps in our understanding still exist, though. More observations are necessary to refine our understanding of how things are changing, whether they might continue to change, and what the consequences of these shifts might be. Should we care about the consequences of changes in the timing of seasonal events? This is the topic for the next chapter.

4

WHY SHOULD WE CARE?

As described in the previous chapter, the changes in the timing of seasonal activity in plants and animals we are experiencing have clear and direct effects on population sizes, food webs, and ecosystem functioning. These changes, however, also have effects that we as humans feel more acutely, directly impacting our health and pocketbooks. As the climate continues to change, these impacts are expected to surface in both expected and surprising ways.

Changes at the Grocery Store

What's your go-to breakfast combo? I typically grab a bagel or bowl of cereal with fruit. And coffee, of course! As the climate changes, it might become harder and harder to reliably enjoy these foods, and changes in plant phenology is one of the reasons.

Agricultural crop losses are one of the most direct ways we experience changing phenology. Crop success is dependent on many factors, including daily temperatures, rainfall, extreme weather events, and the presence of pollinators, pests, and disease, all of which farmers track and use to shape immediate decisions. As climate conditions have changed rapidly in recent decades, farmers have had to adapt. In many parts of the world, farmers are planting earlier in the year. As well, in many places they are harvesting earlier in the year as plants reach maturity more rapidly. For example, some varieties of wine grapes in the north of France are now harvested four weeks earlier than only forty years ago.[1] Sap flow in sugar maples—the key ingredient in maple syrup—has also inched notably earlier in recent years. With continued warming, sap flow is anticipated to shift to an entire month earlier by the end of the century, dramatically affecting the timing of the harvest as well as where sugar maples successfully grow.[2] Likewise in India, the flowering period for saffron has been substantially shortened in recent years. Saffron "threads," highly valued for the flavor and rich color they bring to dishes and drinks, are actually the female reproductive parts of the saffron crocus flower. In parts of India, saffron flowers now open when temperatures are too warm for their development. This leads to a high rate of flower death, and with no flowers, there is no saffron. Between 2013 and 2017, saffron production in Kashmir declined by 90 percent.[3] Consequently, many saffron farmers are shifting their plantings to higher elevations with cooler temperatures.

The northeastern United States is a major fruit production region, as are southern states. As temperatures in these regions have warmed, leaf out and flower bud development as well as the last spring frost have shifted earlier in the season. In many locations, however, the date of the last frost has not shifted earlier to the same degree as plant activity. Consequently, tender plant tissues are at greater risk of exposure to damaging frosts. Many of the plants that produce fruits we enjoy, including blueberries, apples, and cherries, open their flower buds early in the season, sometimes even before they break leaf buds. Once flower buds begin to open, they become sensitive to cold temperatures.

As with saffron, if flower buds are killed by frost, there are no fruits. Early spring emergence followed by a damaging freeze event is occurring with increasing frequency worldwide. Since Thoreau recorded observations of wild blueberry flowers at Walden Pond, Massachusetts, in the 1850s, flowering time has advanced by three weeks, putting the plants at greater risk of frost damage. In 2012, a year with a remarkably early arrival of springtime warmth, blueberries flowered six weeks earlier in Boston than in

Thoreau's time. While these crops seemed to fare reasonably well, others did not. In Michigan that same year, more than 90 percent of the sour cherry crop was lost, totaling $200 million.[4] Spain experienced an 80 percent loss in almond crops in 2022 due to a similar sequence of events: early flowering followed by an extended freeze.[5] And in 2023, blueberries in Mississippi suffered severe damage from subzero temperatures arriving after plants flowered or set fruit.[6] Advancing phenology is expected to worsen the risk of frost damage in the coming decades. The start of springtime biological activity in the United States is projected to advance by up to three weeks by the end of the century.[7] One set of predictions indicates that we can expect to experience early warm springs followed by damaging freeze events in nearly one out of every three years by the mid-twenty-first century. The same is predicted for Europe and Asia, with up to a third of Europe and Asia's forests predicted to be threatened by frost damage in future decades. This paints a gloomy picture for the future of berries on my breakfast cereal.

Warmer winters also negatively impact fruit yield. Blueberries—as well as many fruit and nut trees, including apples, pears, apricots, peaches, pomegranates, plums, walnuts, almonds, and pistachios—require exposure to extended periods of winter chill or cold temperatures to produce large quantities of fruit. This adaptation prevents plants from initiating leaf out or flowering in response to a midwinter warm spell and inadvertently exposing sensitive tissues to cold temperatures. Nevertheless, extremely warm winters can prevent plants from being exposed to sufficient chill, and fruit yield and quality can be severely reduced. In 2017, a year characterized by an extremely early start to spring warmth in the southeastern United States, Georgia suffered an 80 percent loss of peach crops, primarily because the peach trees were not exposed to the cold temperatures necessary for fruit development.[8] To counteract this challenge, growers are exploring a range of strategies to prompt plants to "skip" their need for exposure to chill like removing their leaves after harvest either mechanically or with chemicals and developing cultivars with reduced chill requirements.

More unwelcome news: coffee is also at risk due to changing phenology. Coffee berry crops are most robust and most efficiently harvested when flowers open and pollinated at the same time, resulting in berries ripening at the same time. Increasingly, unpredictable rains are leading to staggered flowering and fruit ripening over the growing season, resulting in inconsistent bean quality and less efficient harvests. In addition, heat stress and drought cause trees to abort flowers, decreasing yields. These concerns, combined with other stressors including increased pests, loss of pollinating insects, and loss of productive lands, paint a less-than-rosy picture for the future of coffee.

An Earlier Start to Flower Festivals?

The subtle but persistent change in the timing of seasonal activity has clear economic implications for tourism and recreation too. The historical timing of many festivals, such as those planned to celebrate lilacs, tulips, or cherry blossoms, no longer reliably coincides with the natural events they honor. Holland, Michigan, hosts the longest-running tulip festival in the United States, boasting more than five million tulips and drawing over six hundred thousand visitors. The organizers have had to adjust the timing and length of the festival to keep pace with the showy but ephemeral blooms as climate conditions in the region have rapidly changed. In the mid-1980s, the organizers extended the festival from four to ten days to accommodate increases in visitors. As spring temperatures steadily increased in subsequent years, however, peak tulip bloom drifted ahead of the event. In 2001, the organizers shifted the event a week earlier and shortened it by two days to better coincide with flowering.

Flower festivals worldwide are being impacted by changing temperatures. In Canada, tulip festival organizers are experimenting with multiple strategies to delay tulips from emerging and blooming in advance of the events, including adding snow cover or heavy mulch, or irrigating. In the United Kingdom, the organizers of Daffodil

The historical timing of festivals, such as those planned to celebrate lilacs, tulips, or cherry blossoms, no longer reliably coincides with the natural events they honor.

Weekend, a major flower festival in Cambridge, have nudged the event earlier by more than three weeks since 1969. And in 2018, the organizers of the National Cherry Blossom Festival in Washington, DC, extended the event to four weeks to maximize the chances that the festivities coincide with peak bloom in thousands of cherry trees.

Cherry blossoms, the showpiece of many flower festivals around the globe, are also experiencing dramatic change. In the United States, peak bloom at the National Cherry Blossom Festival has shifted by nearly six days since it began in 1921, and in Japan, cherry trees are flowering earlier now than at any time in the previous twelve hundred years. The earlier flowering puts tender blossoms at risk of frost damage, and in recent years, blooms have been cut short or ruined altogether by winter conditions following early spring. In 2017, a mid-March frost wiped out at least half of the Yoshino blossoms at the Tidal Basin in Washington, DC. A very early start to spring in 2023 again caused anxiety among National Cherry Blossom Festival planners. Though the festival was scheduled for March 20 to April 16, the indicator tree—a cherry tree that reliably flowers a week to ten days ahead of the other trees in the Tidal Basin—started to bloom in late February. Peak bloom indeed arrived early, though mercifully still fell within the festival window, on March 23.[9]

Shifts in bloom timing are increasingly disrupting cultural activities where plants play a central role. In India, blooms of the laburnum tree feature largely in the Vishu festival celebrating the Hindu vernal equinox in Kerala. Increasingly, the blooms are spent by the time of the festival in mid-April, requiring celebrants to substitute plastic flowers.

Flower festivals can be a significant source of income for host cities, and changes in flowering timing and display stands to have major economic impacts. Increasingly unpredictable bloom timing in such charismatic plants puts strain on event planners and has the potential to substantially reduce tourism dollars brought in by flower events. Similarly, venues hosting destination weddings boasting showy flower displays are dealing with increasing uncertainty in the timing and vigor in blooms. The extremely early springtime warmth in the eastern United States in 2023 demonstrated this concern in concrete terms. Tulips bloomed weeks ahead of schedule in Virginia, spoiling colorful backdrops for weddings later in the season.

At the other end of the growing season, autumn color displays are likewise becoming disrupted and diminished by changing conditions. Primarily due to warmer fall temperatures, leaf senescence has drifted progressively later in the deciduous forests of North America, Europe, and Asia. Drought stress, insects, and disease are also increasingly causing trees to drop leaves before they change color. Leaf peepers—tourists visiting to enjoy fall colors—are estimated to contribute over $30 billion annually to the economies of the eastern US states.[10] The disheartening trend of diminishing fall color displays stands to have substantial negative impacts on local economies.

It's not just changes in plant activity that affect tourism and recreation; changes in animal phenology are affecting tourism economies as well. An increasingly later arrival of migratory sardines to the eastern shores of South Africa is likewise impacting tourism and recreation to the region. The appearance of these massive schools of small fish—which can measure several kilometers long—boost fishing conditions as well as shark, dolphin, and penguin activity. Since the middle of the previous century, this event has drifted later in autumn by over a week in response to changes in sea surface temperature. Continued changes to the timing of this seasonal event have the potential to further impact both fishing and tourism as well as disrupt predator-prey relationships, as they are an important food source for sharks.

Longer Growing Seasons and Your Health

Achoo! Sniffle. Does it seem like your allergies have gotten worse lately? It sure feels like it to me. If you said yes, you're not wrong. Since 1990, the pollen season has lengthened by over twenty days in the United States and the amount of pollen generated by wind-pollinated plants has increased by about 20 percent.[11] Why is this happening? The higher atmospheric carbon dioxide concentrations and warmer global temperatures we are now experiencing essentially fertilize plants, enabling them to grow larger and produce more pollen than in the past. The consequences for those of us who suffer from seasonal allergies are real. Dr. Stanley Fineman, an allergist who has practiced in Atlanta, Georgia, for over forty years, recently shared that in years past, he would instruct his patients to begin using their allergy medications on Saint Patrick's Day (March 17) to be prepared for upcoming surges in airborne pollen. Yet because the growing season—and consequent pollen production—starts so much earlier now, he has changed his advice to initiate medication on Valentine's Day (February 14), a full month earlier.[12]

Changing climate conditions also benefit many insects that carry disease. Milder winters and a longer growing season encourage earlier emergence and later fall activity in ticks, which carry Lyme disease and other pathogens, and mosquitoes, which carry dengue, West Nile virus, malaria, and multiple forms of encephalitis. The longer seasons of insect activity increase the period during which humans can be exposed to these disease vectors. The same is true for waterborne illnesses: warmer fresh and marine waters promote growth in harmful algae, bacteria, viruses, and parasites, and likewise increase the potential for exposure through a longer period of activity.

Since 1990, the pollen season has lengthened by over twenty days in the United States.

Longer Growing Seasons Affect Carbon Storage

The facts we learned in elementary school are correct: plants "breathe in" carbon dioxide, and in combination with sunlight and water, produce carbohydrates—the building blocks of tissues—and oxygen. With more carbon dioxide in the air, plants grow larger and more rapidly; the additional carbon dioxide effectively acts as a fertilizer. In many parts of the world, the extended growing season is, indeed, allowing plants to grow larger and longer, resulting in increased storage of carbon in plant tissues.

Increased carbon storage conferred by a longer growing season sounds like a fantastic solution for pulling carbon out of the atmosphere. The idea to plant millions of trees to address climate change recently popularized by social media influencers, companies, and even some governmental agencies is inspired by this logic. The proposal is to plant more trees, and as they grow, they will sequester carbon in their tissues and the soil. Unfortunately, there are limits to how far this can go toward offsetting the continually increasing carbon dioxide in the atmosphere. In addition to carbon dioxide, plants need water and nutrients to grow, and both ingredients limit how many additional trees landscapes can support, especially in water-limited parts of the world.

The gains in carbon storage won through an earlier start to the growing season likewise have limits. Advanced leaf out and flowering puts plants at greater risk of subsequent frosts, which can lead to carbon losses through tissue damage. In addition, earlier leaf out has been tied to earlier senescence—the breakdown of leaf tissue at the end of the growing season—in some forests, which similarly results in no overall gains in carbon storage. Further, it might surprise you to know that plants release some of the carbon dioxide they take up through respiration. In a warmer world, the amount of carbon dioxide they release through respiration is anticipated to become much greater, potentially completely offsetting carbon uptake achieved through additional plant biomass.

So then, how much can an extended growing season help us address the problem of continually increasing carbon dioxide in the atmosphere? The answer remains unclear, and researchers are actively pursuing this question. A recent study using data from across the northeastern United States indicated that spring-flowering herbs—the shorter-lived plants that occupy the forest floor—are advancing their phenology more rapidly than the longer-lived, overstory canopy trees. The different rates of advancement in activity among these groups offer the herbs a longer window with access to light before the canopy closes above them, enabling increased seed production, growth, and associated carbon storage among these plants.[13] Another study addressing the same question, however, suggests an alternative story: that overstory trees are advancing the timing of their leaf out more rapidly than the understory plants, thereby reducing the herbs' access to light, and curtailing their ability to flower and produce viable seeds.[14] The authors of this second study estimate that the reduction in herb growth resulting from these changes could result in a substantial *decrease* in the carbon stored by these plants by the end of the century.

Though the studies' conclusions are conflicting, the underlying explanation for the different rates of change is the same: the overstory trees and understory herbs are cued to leaf out by different environmental conditions. In the northeastern United States, where both studies took place, leaf out in trees is generally governed by air temperature, whereas the understory herbs pay attention to snow depth and soil temperature to time their activity. Interestingly, in Asia and Europe, there seems to be no divergence in the timing of activity between canopy trees and understory herbs, as neither plant group demonstrates strong temperature sensitivity on these continents. Consequently, we shouldn't expect increases or reductions in carbon storage among understory herbs in these regions, at least not because of different rates of change in the timing of their leaf out. This is just one example of how ecosystem carbon cycling stands to be dramatically altered by changes in plant phenology and illustrates the difficulty in anticipating how things might change in the coming decades; clearly, there is much more work to be done.

Large wildfires also dramatically affect carbon sequestration, releasing massive amounts of carbon into the atmosphere. Longer growing seasons, characterized by warm spring temperatures earlier in the year, drive increased growth in annual and herbaceous perennial plants. Lushness in fast-growing plants, including annual grasses, can provide ample fuel to sustain fires. This is especially true in the western United States, where introduced grasses such as cheatgrass, buffel grass, red brome, cogongrass, and several others are increasingly widespread. The wildfire season in the United States is now nearly three months longer than fifty years ago due to a warmer and longer growing season.[15] This pattern extends to other regions as well; in Australia, the season has lengthened by nearly a month in the past four decades.[16] And with more frequent, larger fires, more carbon is being released into the atmosphere.[17]

Fires, in turn, impact phenology. Large, hot fires can lead to changes in the dominant vegetation in a region, often from forests to shrubs and young trees. These changes in vegetation can lead to changes at both the start and end of the growing season. Recent fires in the western United States have led to earlier and longer growing seasons. The costs of controlling these wildfires have also dramatically increased; in the United States alone, wildfire suppression costs exceeded $3 billion in 2018.[18] The proportion of the US Forest Service's budget that is allocated to managing wildfire has shifted from 16 percent in 1995 to more than 50 percent in 2015, and is projected to continue to increase as global temperatures warm further.[19]

Changing Phenology Disrupts Harmony among Species

An especially troubling consequence of shifting phenology is the disruption of synchronicities between interacting species such as plants and the insects or animals that pollinate them. Interactions between species are at risk of disruption when their seasonal events are cued by different environmental drivers. For example, springtime activity in many plants is cued by exposure to warmth, and in warm years, these plants leaf out and bloom earlier. In contrast, many insects use daylength as a cue to emerge from winter diapause and so do not become active earlier in years with warmer spring temperatures. Such differences mean that plants and their pollinators will be active at the same time in some years and less so in others. As temperatures continue to increase, the likelihood of mismatches like this emerging increases. This creates major problems when species depend on one another for resources or services, and can have effects that extend to humans.

When plants and their pollinators become mismatched, no longer flowering and flying at the same time, both parties suffer: flowers lose their pollination services—necessary for fruit and seed development—and pollinators lose their source of food. Plant-pollinator relationships are being disrupted around the globe, though how things are changing varies among species pairs. For example, in northern Japan, flowers are blooming before bees emerge, resulting in a decrease in seed production. In Spain, the opposite is happening: insects are advancing their activity more rapidly than plants. But the outcomes are the same: both flowers and insects are losing out on critical resources and benefits.

In other systems, insects and the plants they pollinate are shifting the timing of their activity at similar rates, thereby not becoming decoupled. For instance in Wisconsin, United States, wild blue phlox, red columbine, and Canada anemone have all advanced their flowering time in recent decades. These species also show increased visits to flowers by the bees, beetles, and butterflies responsible for pollinating them, suggesting that these insects are shifting their activity alongside the flowers.[20] It appears that activity in both the plants and insects is cued by the same variable: temperature. Unfortunately, for many plants and insects, we don't know the specific cues to initiate activity in the spring. Consequently, our ability to anticipate future mismatch among plants and pollinators is quite limited.

In marine communities, predator-prey relationships between fish and the plankton—tiny waterborne plants

When plants and their pollinators become mismatched, both parties suffer: plants lose their pollination services, and pollinators lose their source of food.

or animals—on which they feed are likewise becoming mismatched. Plankton growth appears to be cued by increasingly available light as days lengthen in the spring, whereas fish egg hatch is cued by water temperature. As water temperatures have increased, fish development has gotten earlier. Unfortunately, plankton blooming is not shifting earlier in the spring, and fish are going hungry. In the English Channel, for example, young fish development has advanced by twelve weeks over the last twenty-five years as water temperatures have rapidly increased. In contrast, plankton blooms have occurred at essentially the same time each spring in response to lengthening days.[21] In other situations, plankton blooms have shifted much earlier, no longer coinciding with fish hatch and development. Such mismatches have the potential to dramatically impact the shape of ocean ecosystems, the fishing industry, and the fish we eat.

Photos of polar bears stranded on ice floes, footage of glaciers calving, and time-lapse maps of glacier retreat: the Arctic has long been an icon of climate change. Evidence of emerging mismatches is piling up in the Arctic, where rates of warming and consequent changes in phenology are among the most rapid. Arctic snow geese are poised to suffer steep declines in numbers due to mismatches between the timing of egg hatch and peak forage amount and quality. Despite rapidly warming springtime temperatures that are driving clear advancements in grass maturation, geese have not advanced their nesting dates, as doing so risks the eggs being eaten by polar bears. Arctic caribou are similarly showing signs of declines in number due to mismatches. Plants eaten by caribou mothers and calves have rapidly advanced their growing season, whereas the caribou have not shifted the timing of their calving. Over a period of less than twenty years, offspring survival declined by a staggering 75 percent.[22] Snowshoe hares are especially vulnerable. The hares' coat color shifts from winter white to brown in the spring, at close to the same time from one year to the next. Yet increasing spring temperatures have pushed snowmelt progressively earlier, leaving the still-white hares on a landscape devoid of snow and thus exposed to predators. Researchers estimate that if hares cannot quickly adapt to the rapidly changing conditions, they risk a color mismatch with their environment upward of ten weeks by the end of the century, likely spelling major trouble for their population numbers.

Perhaps surprisingly, some interacting species are shifting their phenologies in tandem, with no major disruption to their interaction. For instance, Aldo Leopold, the conservationist credited with pioneering the field of wildlife ecology in the mid-twentieth century, documented first bloom in skunk cabbage at his shack in southern Wisconsin around the first of April between 1935 and 1945. He reported eastern phoebes, which overwinter in the southeastern United States, first arriving about a week later. The phoebes hang out around skunk cabbages for food; the malodorous plant attracts insects that the birds eat. In recent years, skunk cabbages have bloomed in mid-March—weeks earlier than in Leopold's time. The phoebes have similarly advanced their arrival, appearing in recent years around March 24, and continue to take advantage of skunk cabbage's ability to attract insect food, just as in Leopold's time.[23] While the exact reason this relationship has remained intact despite shifting earlier is not entirely clear, a likely explanation is that the birds and plants both use temperature as a cue to initiate springtime activity.

Disrupted interactions between plants and animals also directly impact traditional culture and ways of life. Elders of the Confederated Tribes of Siletz Indians recently shared that the appearance of carpenter ants historically indicated the time for catching eels. This cue is no longer meaningful because recent changes to the weather have disrupted these relationships.[24] The rapidly changing climate is disrupting this relationship and the tribe's activities. Around the globe, many Indigenous communities use ecological calendars comprised of both physical events such as the first snowfall and biological events such as migratory fish arrival. Rapid environmental change is rendering these calendars less reliable. Further, rapidly changing seasonality risks long-lasting cultural disturbances and negative impacts to well-being. Dr. Samantha Chisholm Hatfield, assistant professor at Oregon State

University and enrolled member of the Siletz tribe, shares that disruptions and mismatches are not just an inconvenience but also "challenge the fundamental belief about how elements of the natural world are connected."[25]

Many invasive species are benefiting from changing climate conditions too. Invasive species are plants or animals that once introduced to a novel ecosystem, become abundant on the landscape and cause disruptions. Invasive plants are frequently superior competitors for multiple reasons, one of which is the ability to leaf out or flower ahead of native plants, enjoying priority access to nutrients and light. These plants also frequently flower longer and retain their leaves later into the autumn, enhancing further growth. As well, many successful invaders exhibit a greater capacity to shift their leaf out or flowering times in response to rapidly changing climate conditions than native plants, suggesting that warming temperatures may promote further invasions. As an example, Japanese barberry was introduced to the United States as an ornamental plant in the 1800s. Since that time, the shrub has spread widely throughout the central and eastern United States and Canada. Because barberry leafs out earlier in the spring than native shrubs, it outcompetes them for light and nutrients. Across the United States, forest understories have been transformed from a mix of native shrubs to dense thickets of barberry that provide habitat for disease-carrying ticks.

Downstream Impacts of Phenological Mismatches

The impacts of changes in phenology—and consequent mismatches—can cascade through ecosystems. The most visible impacts are declines in species numbers for one or both members of a species interaction. Disrupted interactions, however, affect other aspects of ecosystems that are not as visible, such as water and nutrient cycling, and can indirectly impact humans.

As an example, the emerging mismatch between migratory birds and their caterpillar food is changing the storage of carbon and other nutrients. As described previously, many migratory birds are now arriving at their summer breeding grounds after most caterpillars—their primary food source—have emerged. The delays in bird arrivals is giving caterpillars more opportunity to consume plant tissue. Extended periods of caterpillars munching on leaves reduce the amount carbon stored in tree tissue, as it instead passes through the caterpillars' guts and then back out again.

Not all mismatches lead to reduced nutrient availability and carbon uptake. For instance, the delayed arrival of migratory caribou to inland Greenland provides the plants they consume more time to grow and take up carbon and other nutrients.[26] The long-term implications of this sort of ecosystem shift are yet to be seen, though could result in changes in plant community composition.

Changing phenology, while seemingly innocuous, impacts our lives in many clear and tangible ways.

In large part, the downstream impacts of species mismatches have not yet been directly evaluated. Additional investigations are needed to better understand and predict their implications, especially because with additional global temperature increases and the resultant additional phenological shifts, the interactions and consequences will likely change further.

In mid-December 2023, my friend and colleague Dr. Jorge Santiago-Blay sent me photos of cherry trees on the Penn State York campus in Pennsylvania, United States, bearing many open blooms. While he expressed pleasure over the lovely sight, the phenomenon sparked confusion and disorientation because typically, eastern Pennsylvania is wracked with frigid temperatures and snow by this time of the year—hardly favorable conditions for delicate cherry blossoms. With increasing variability and change in the earth's climate, we are poised to experience oddball phenological events such as this with increasing regularity, with impacts to not only to the species exhibiting the unseasonal activity but also our health, cultures, diets, and mental states. Changing phenology, while seemingly innocuous, impacts our lives in many clear and tangible ways.

5

PUTTING OUR UNDERSTANDING OF PHENOLOGY TO WORK

As the days lengthen and temperatures warm in the spring, those of us who live in seasonal climates respond by shedding our heavy coats and sweaters, donning sunglasses, taking longer walks, and dining outdoors. Likewise, the shorter, cooler days of autumn and winter prompt us to pile layers back on and pivot to indoor activities. In the same way, plants and animals undergo changes as days get longer and warmer, and then again as they shorten and cool off. The environmental conditions that prompt such changes are referred to as drivers or *cues*.

Cues: The Conditions That Cause Seasonal Events to Occur

The specific conditions that trigger plants and animals to wake up in the spring and conclude activity in the autumn are different from one species to another. The three most influential variables, especially for plants, are temperature, daylength, and available moisture. The importance of these factors varies around the globe. Temperature and daylength play the largest roles in the temperate and boreal forests of middle to high latitudes as well as elevations where plants experience clear periods of winter dormancy. Moisture and daylength play larger roles at lower latitudes—nearer to the equator—where temperatures don't get cold enough to limit growth and activity in plants and animals.

In temperate and boreal ecosystems, many of the plants that leaf out or flower earliest in the season, like lilacs, hazelnuts, poplars, and birches, need only for temperatures to warm sufficiently to trigger them to leaf out. Many short-lived, herbaceous plants similarly rely solely on warmth to trigger spring growth. The strategy of only requiring exposure to warm temperatures to trigger activity employed by these plants enables them to take advantage of years with an early start to spring, extending their growing season and gaining a competitive advantage over plants initiating their activity later in the season.

Other plants are more conservative in timing their spring season activity. Many plants that leaf out later in the season, such as hackberry and oaks, only acknowledge springtime warmth once days have exceeded a required number of hours of daylight. This prevents them from

The three most influential variables cuing phenological events, especially in plants, are temperatures, daylength, and available moisture.

accidentally exposing their tender plant tissues during the winter months in response to an anomalous warm period. A drawback of relying on a daylength cue is the inability to take advantage of very early springtime warmth. If warmth occurs prior to these plants' daylength requirements being met, the plants do not respond. This risk-avoidant strategy is common among long-lived plants in stable plant communities, enabling them to persist through variable conditions for many decades.

In a third strategy, plants largely ignore springtime warmth until they have been exposed to sufficiently cold temperatures or a certain amount of "chill" during the winter months. This approach similarly protects plants from being tricked out of dormancy by midwinter warm days. Many plants in temperate environments, especially long-lived ones wishing to minimize the risk of frost damage, have both daylength and accumulated chill requirements necessary to cue growth in the spring.[1]

In tropical environments, where temperatures remain above freezing year-round, and seasons are defined as either "wet" or "dry," rainfall plays a greater role in governing phenology. Seasonal fluctuations in temperature and daylength are less dramatic in these ecosystems than at higher latitudes. The same is true for desert and semiarid systems. In these dry climates, precipitation plays the largest role in cuing seasonal events, while temperature and daylength also frequently matter. In the Sonoran Desert of Arizona, California, and northern Mexico, plants from the low desert to pine forests withhold summertime flowering until the summer monsoon storms commence.[2] In colder, higher-elevation montane and alpine environments, the timing of spring snowmelt is the most important factor in shaping the timing of plant growth. Daylength and temperature again play a secondary role, preventing plants from initiating activity in response to early snowmelt when the risk of frost damage is high.

Temperature is a major factor driving springtime activity in many animals as well, including the emergence of many insects from winter diapause, amphibian emergence from winter brumation, fish egg hatch, and emergence from hibernation in small mammals such as squirrels and bats as well as larger mammals like bears and marmots. Temperature also plays a big role in the timing of bird activity, especially in short-distance migratory birds, cuing them to initiate spring and autumn migrations. In addition, some later-season activities, such as breeding in amphibians, are controlled mainly by temperature. In aquatic systems, water temperature is frequently the primary factor shaping egg hatch and migration.

For animals that migrate long distances, such as long-distance migratory birds and monarch butterflies, photoperiod—the number of hours of daylight in a twenty-four-hour period—is the primary factor shaping the timing of their departure. Increasingly longer days prompt these species to begin their journeys. Photoperiod is important in initiating and ending winter diapause in many insects too.

Species Also Influence Each Another

You know that tendency we humans have of looking around to see what other people are doing and using this information to shape our own actions? Plants and animals do the same thing, and this proclivity can shape when species undergo seasonal transitions in the present as well as in future generations.

As an example, the flowering time in plants that rely on insects, birds, or bats for pollination is shaped not only by temperature and daylength but by when pollinators are active too. How is this so? Let's back up to why plants bother with the effort of producing flowers. The main reason plants produce flowers is to develop fruits and seeds that germinate and grow into new baby plants. Most species require pollen from other individual plants to successfully produce fruits and seeds. Pollinators like birds and insects perform this pollination "service" for plants, unintentionally moving pollen among plants as they forage for nectar.

Through the service of moving pollen from one plant to another, pollinators indirectly shape when plants flower. Plants that flower when pollinators are active are the ones that receive pollen and successfully produce seeds. The subsequent generation of plants originating from these seeds is likely to flower at the same time of the season as its parent plants since the genetic material controlling for flowering at that time is passed on in the seeds. In contrast, plants flowering at times other than when pollinators are present are less likely to be pollinated and produce seeds; consequently, their genes are not carried forward in seeds, and flowering at these "off" times is thereby deprioritized.

The introduction of novel species into a community can also shape when various members of that plant community leaf out or flower in an indirect way. By creating competition for light and soil nutrients, these newcomers force the plants that were there already to either shift their activity earlier in the season or risk missing out on resources. Likewise, herbivory—when animals and insects feed on plants' leaves, flowers, or seeds—can shape plant phenology. Insects, which damage leaf and shoot tissues at a cellular level when they consume plant material, can cause trees to drop leaves earlier than they otherwise would. Ungulates like deer tend to remove entire leaves or branches as they browse. Their munching can push leaf senescence and leaf drop to occur later into the season.

Using Seasonal Activity as an Indicator

"Plant corn when oak leaves are the size of a squirrel's ear." "Protect apples from maggots when Canada thistle is in bloom." Perhaps you've heard adages like these, which are frequently repeated by university extension professionals, horticultural and landscape blogs, and even the *Farmer's Almanac*, though many of these guidelines have their origins within Indigenous communities. Many Indigenous cultures have lengthy traditions of using indicators to cue activities and have generously shared this knowledge. For example, Lil'wat elder Nellie Wallace Peters shared that when wild roses bloom on the Lil'wat Nation in British Columbia, Canada, it is time to harvest basket grass and cedar roots.[3] Likewise, Seine River First Nation knowledge holders have shared that the presence of Tiger Swallowtail butterflies in Ontario indicates peak sturgeon spawning.[4] As well, Nlaka'pamux knowledge holders shared that wild rose flowering indicates that soapberries are ready to harvest.[5] The use of indicators extends to agricultural practices in other cultures as well. For instance, farmers in Montana, United States, historically used lilac flowering as a cue to cut alfalfa to minimize damage from alfalfa weevils, which overwinter as eggs and hatch after temperatures warm sufficiently.[6]

These adages and indictors perform well in many cases because the species of focus share similar cues to their seasonal events. The Canada thistle and apple example employs thermal sums in an indirect way. Canada thistle flowers in late spring and early summer, once soil temperatures reach 35–45°F. Similarly, apple maggot adults emerge and begin to fly after nine hundred growing degree days have accumulated (base 50°F), which also occurs in late spring and early summer in the upper midwestern United States. Accordingly, apple growers can use other plants—Canada thistle, in this case—as a reasonably reliable indicator of when to protect their fruit crops to minimize damage.

Indicators such as these can be established through careful observation and recordkeeping. Dr. Dan Herms, professor of entomology at Ohio State University, collected observations of flowering and pest insect activity for several years at arboretums in Ohio and Michigan. As Herms describes, once the sequence of activity in flowering plants and troublesome insects has been established at a location, these events can be used as indicators for when to act. As an illustration, pine needle scale is a sap-feeding insect native to the United States that can kill branches or even entire trees. Herms observed that the first generation of pine needle scale eggs hatch at approximately the same time that flowering dogwood, red chokecherry, and Vanhoutte spirea begin to flower. As such, horticulturalists might use flowering in these plants to indicate when to control the insect. Or better yet, a landscaper might keep watch for first bloom in common lilac, Ohio buckeye, or Tatarian honeysuckle, all of which begin to bloom about a week prior to the hatch of the first generation of pine needle scale, according to Herms's observations. Flowering in these plants can be a signal to managers to get ready to take action very soon.[7]

Using Cues to Predict Activity in Plants and Animals

Who doesn't wish they had a crystal ball that could tell them when things in the future are going to occur? A thorough understanding of the environmental conditions that cue phenological transitions can get us close, at least for some things. If we know the conditions an organism requires to undergo an event like leaf out or egg hatch, we can predict when the event is likely to occur by tracking the associated conditions.

Accumulated springtime warmth has been used to predict the timing of activity in plants and insects for nearly three hundred years. The approach of tracking warmth was first described in the mid-eighteenth century by René de Réaumur, who proposed that plants reached various stages of development after being exposed to sufficient warmth.[8] Using accumulated warmth to predict development in organisms essentially involves comparing the daily average temperature at a location with a base temperature,

If we know the conditions an organism requires to undergo an event like leaf out or egg hatch, we can predict when the event is likely to occur by tracking the associated conditions.

below which the plant or animal of interest is biologically inactive. When the average temperature on a day exceeds the base temperature, the difference between these values is the amount of warmth, measured in thermal units such as degree days, accrued on that day. Thermal units are accumulated over the spring, often beginning on January 1 in the Northern Hemisphere, by adding the thermal units accrued on a day to those gained on every previous day following the selected start date.

The specifics of calculating thermal units, including the choice of base temperature, start date for accumulating warmth, and method for calculating thermal units within a twenty-four-hour period have been debated, tweaked, and refined since Réaumur's original approach, which involved the use of his custom thermometer that scaled from 0° for the freezing point of water to 80° for the boiling point.[9] Because the approach generally performs quite well at estimating activity in particular organisms, it has been widely implemented in agriculture, horticulture, and pest management. As an example, corn generally emerges from the ground once 100 growing degree days have been accumulated after planting, and then requires 2,700 growing degree days until ready to harvest.[10] Growers can track growing degree day accumulations on their farms to estimate harvest dates and optimize efforts.

Thermal sums are also used in pest management applications to schedule control activities to maximize efficacy.

For example, emerald ash borer is a beetle native to Asia that was accidentally introduced to the United States in 2002. In their larval stage, the beetles destroy ash trees by tunneling into the tissue that moves water and nutrients, killing them within a few years. Since their introduction to the United States, emerald ash borer beetles have spread unchecked across the eastern and central parts of the country, killing hundreds of millions of ash trees. Because the beetles spend most of their lives under the bark of ash trees, they are difficult to control. Once they reach their adult stage, however, they exit trees with the hope of finding a mate, and in the case of females, laying eggs. This window, which lasts for approximately two weeks, is a prime opportunity to limit their spread to new trees by trapping the flying beetles. Adult emerald ash borers emerge after host trees have been exposed to a particular amount of warmth in the spring, specifically 450 growing degree days (base 50°F).[11] Pest managers can track the accumulation of growing degree days at locations they manage to estimate the period when adult emerald ash borers will be in flight.

Winston Churchill is frequently credited with having said, "The farther back you can look, the farther forward you are likely to see," and this insight applies here. If we have knowledge of when things happened in the past, we can use that to predict when they might occur in the future. Making accurate predictions of when plants and animals will undergo seasonal transitions such as leaf out or egg hatch requires an understanding of the specific conditions that cue the event to occur. How can we determine the conditions that cue seasonal events to occur in plants and animals? One approach involves looking for relationships—or *correlations*—between when seasonal events occur and recent, local weather conditions. This approach to estimating phenological cues is relatively straightforward, mainly requiring lots of accurate observations of the event and analytic skill to construct a *model*—a mathematical description of the specific conditions associated with the onset of phenological event.

As an example, Herms determined that black cherry first flowers after 357 growing degree days (base 50°F) have accumulated. He made this determination by observing the flowering time in four individual black cherry trees over three years at the Secrest Arboretum in Wooster, Ohio. Using measurements of daily minimum and maximum temperatures he collected at the site, he calculated how many growing degree days had been accumulated on the day that each cherry tree first flowered in each year of his study. Using these simple calculations, he arrived at the conclusion that these trees require exposure to approximately 357 growing degree day units to bloom. Herms can now estimate when his cherry trees are likely to flower by tracking daily temperatures, calculating how many growing degree days have accumulated, and comparing that number to the 357 growing degree days he knows the trees require to bloom.

The primary drawback of correlative approaches is that they do not unequivocally establish the conditions that cause the event in the plant or animal to occur. We often see earlier leaf out in many plants when springtime temperatures are warmer. But is springtime warmth the sole factor causing the plants to flush their leaves? In temperate ecosystems, spring temperatures increase at the same time that days lengthen. So is it that the plants are responding to warmer temperatures or longer days, or a combination of both factors? What about the preceding winter? Does it matter whether temperatures were cold? By only observing environmental conditions and patterns in plant or animal response, it's tough to say. Yet even though correlative approaches do not spell out the distinct factors that drive activity in organisms, they can frequently be useful, enabling us to predict the timing of activity with acceptable accuracy.

An alternative approach to establish the cues to phenological events is through controlled experiments. Often carried out in growth chambers or laboratories where conditions can be precisely regulated, factors such as daylength and temperature are manipulated to isolate their specific influence on plant or animal response.

Experiments are also often undertaken outside, where organisms are naturally found. While it's much more difficult to control all the conditions that could influence phenology outdoors, this approach offers the benefits of yielding more "real-world" results by inherently incorporating true-to-life environmental variation. Researchers have devised all manner of clever approaches to manipulating temperature, rainfall, snow cover, and other important factors in outdoor experimental plots, including heat lamps, rain out shelters, the addition of nutrients or carbon dioxide, and even adding or removing snow by shoveling it onto or off experimental plots. Field experiments focused on animal phenology are less common, mainly because animals move around, making them more difficult to control. Nevertheless, recent studies have explored the impacts of warmer water and changing light levels on aquatic species as well as increased air temperature on insect phenology and reproduction. The information acquired through these experiments can be applied in forecasting applications.

Forecasting Phenology across Regions

Phenology models—regardless of their origin—can be used to make predictions of when events are expected to occur and discrete locations, such as Herms did in Wooster. You can do this at your home by tracking daily temperatures and calculating accumulated growing degree days using the simple growing degree days method. Many online calculators are available for estimating growing degree day accumulations at a location in real time.

Predictions of when phenological events are expected to occur can be made across geographic regions too, using maps reflecting the cues of the event. For instance, Herms's black cherry flowering model can be used to forecast when black cherry trees are expected to flower at locations across the country using maps of daily minimum and maximum temperature. Such maps, which are readily available for the whole country, are created using temperature observations collected by thousands of weather stations across the country. The maps are "gridded," meaning that the country is divided up into many tiny pixels, like in a digital photograph. Each pixel on the map has a temperature value assigned to it, generated by using mathematical equations to estimate temperatures at locations between the stations. Using these maps, the number of growing degree days that have been accumulated can be calculated for each pixel of the map in the same way that growing degree days are calculated at a single location: by comparing daily average temperatures to the selected base temperature.

Springtime warmth arrived remarkably early in much of the southern and eastern United States in 2023, and plants and animals responded with remarkably early activity. Pecan trees leafed out and blueberries flowered weeks ahead of schedule in Georgia, and wine grapes began to develop fruits in early March in Mississippi, putting these valuable crops at risk of freezing events. Groundhogs in Maine emerged from hibernation weeks early, and tundra swans were observed midmigration in Pennsylvania a month ahead of schedule. Rapid development in cherry trees at the nation's capital caused organizers of the National Cherry Blossom Festival to fret that the blooms would be spent by the time of the event.

To contextualize these anomalous events, the news media heavily referenced "start of spring" maps offered by the USA National Phenology Network (USA-NPN). These maps indicate when spring season activity in plants and animals began, as reflected by recent temperature conditions. A set of companion spring anomaly maps give a sense of whether these conditions occurred earlier or later than what is typical for a location. The anomaly maps are created by comparing when conditions associated with the start of springtime biological activity were reached in the current year to when these conditions are usually reached—specifically, when spring started, on average, in the last three decades. The USA-NPN's maps showed that in 2023, the start of spring arrived two to four weeks ahead of schedule across huge swaths of the southern and eastern portions of the lower forty-eight US states.[12] In contrast, nearly the entire western half of the country experienced a later than average start to spring. These maps helped people across the country to understand just how early or late conditions were at their locations, indicating whether the beginning of the allergy season, appearance of turf pests, and start of the planting season would be ahead or behind schedule that year.

Invoking the Crystal Ball: Making Long-Lead Phenology Forecasts

Weather forecast maps—indicating predictions of weather conditions in the days or weeks to come—can be used to make longer-lead forecasts of when phenological events are expected to occur. The aim of such forecasts is to provide advance warning of when key seasonal events such as egg hatch or adult emergence in insect pests or green up in invasive grasses are expected to occur. While the value of these products is clear, few operational products exist, and this is an active area of research. In one of the first efforts along these lines, a research team based at Cornell University in Ithaca, New York, recently unveiled long-lead forecasts for the start of springtime activity. These maps are first available in December of the preceding year and predict when conditions that trigger leaf out in early spring plants like lilacs will be reached.[13] As spring unfolds, the forecasts are updated to reflect recent conditions. The aim with products like this is to give growers, resource managers, and others a sense of where springtime conditions will be earlier or later than normal with more advance warning.

Predictions of when phenological events will occur can be made even farther into the future—years or even decades. Researchers around the world are doing this very thing to estimate how much phenology might further change in particular species. In Korea, it is estimated that by the end of the twenty-first century, rosebay and forsythia, along with plum, cherry, peach, and apple trees will flower up to two weeks earlier.[14] Similarly, leaf out is predicted to advance by one to two weeks in several common trees including birch, lime, oak, and maple in central Lithuania as well as among many species of coniferous and broadleaved trees in both France and China.[15] Green up in rangelands in the central United States are also expected to advance by ten to nearly twenty days.[16] In some cases, forecasts indicate even more dramatic shifts. Flowering and fruit ripening in several culturally important food-producing plants in the Pacific Northwest region of the United States are predicted to advance by over a month by the year 2100, with the largest advances anticipated at higher latitudes and elevations.[17] Fewer studies predict changes to the end of the growing season, though multiple studies indicate an earlier end to the growing season in both Asia and the United States.[18] Overall, however, the research is quite spotty, and we are far from having a complete picture of what the future might bring.

Our Ability to Predict the Future Is Limited

Evidence is mounting worldwide that phenology is changing, and our understanding of why—the underlying drivers of phenology and how they are being impacted by changes in our climate—is steadily growing. Even so, our ability to predict changes in phenology and the consequences of this change for the plants and animals that will make up communities along with their ability to carry out critical ecosystem functions is still woefully inadequate. So much remains to learn about the specific cues to phenological events in individual species, whether individual species have limits to their ability to shift phenology, and how changes might interact with the environment to elicit further changes. Findings from one species or system don't necessarily translate to others. In short, we need a lot more observations of a lot more species collected in a lot more locations to truly understand how things are changing, and what those changes might mean for the health of our ecosystems and ourselves.

6

YOU CAN PLAY A ROLE TOO

We are living in a period where "not normal" is normal. As I write this, in summer 2023, record-setting wildfires are burning out of control in nearly every Canadian province, and the smoke is choking much of Canada and large swaths of the eastern United States. Just three years ago, wildfires burned for weeks in the mountains just north of Tucson, where I live. At night we watched the flames engulf individual ponderosa pine trees from our backyard. At times it's hard to not feel immobilized by despair.

Also happening as I write this: a mother broad-billed hummingbird is incubating two mini jellybean-sized eggs in a nest made of fluff and grass a mere five feet from my desk. As I work, mama hummer whizzes from the nest down to the yellow bells blooms for a sip of nectar and then back to the nest, then over to the desert honeysuckle for another snack, and then back to the nest. This is the third year in a row that this creature has nested on our back porch, returning to our backyard to build her nest precariously on a string of twinkle lights just feet from the screen door that slams shut with a loud *bang*. I cannot fathom her choice of nest location, though I am forever grateful, because watching baby hummingbirds fledge is nothing short of magical. At times it's hard not to feel bowled over by wonder.

Over the last 150 years, the world's climate has changed noticeably. A century and a half ago, 110°F days were rare, occurring only once or twice a season here in Tucson. This summer, we are setting records for daily high temperatures and sustained extremely hot conditions, with 17 days overtopping 110°F. The news is full of reports of our iconic saguaro cacti dropping arms and beehives melting in Phoenix, where it has been exceptionally hot. And this record-hot year follows on the heels of many other record-hot years: 8 of the past 20 years have been the warmest on record in Arizona. The impacts on hummingbirds as well as other plants, animals, and humans are becoming more apparent each year.

In particular, our mental health is taking a hit. Anxiety and depression related to climate change and its consequences are increasingly common maladies. Google searches for "climate anxiety" increased fivefold in 2021, and mental health experts are fielding a rapidly growing number of requests for help dealing with eco-anxiety, eco-grief, and climate doom.[1] It is far too big of a problem for any of us to address alone. Timothy Morton, an environmental philosopher, has labeled climate change a "hyperobject"—something too big to comprehend or assimilate into our individual, day-to-day existences.[2] Evidence of rapid and unprecedented change surrounds us, and the news is persistently and painfully gloomy and doomy. The impacts of changes we can make on a personal level feel like a raindrop in an ocean-sized problem. I often find myself either frustrated and disappointed that I'm not doing more, or disheartened and drained by the constant drumbeat of bad news followed by more bad news.

Yet watching mama hummingbird tend to her babies has taught me that focusing on the small wonders unfolding in my midst calms and refocuses my scattered and anxious thoughts. My routine of circling the yard to collect phenology observations, originally established to generate records of seasonal phenomena, brings me so much more. Pausing to witness leaves unfurling and insects' dogged activity serves to reframe the world for me. This shift in perspective, offering relief from climate anxiety, is accessible to everyone. Just about everyone understands "blooming" and "mating," and anyone can document these phenomena—no formal training or specialized equipment necessary.

Everyone Can Play a Role in Documenting Phenology

When our first son, Zachary, was born in 2008, my husband, Mike, and I planted a desert willow tree in our backyard. In the intervening fifteen years, the tree has gotten huge—much larger than most other desert willows around town. My guess is that over the years, our tree has learned how to take advantage of our aging sewer pipes; desert plants have to be crafty to survive.

I love this tree. Its branches overtop our small backyard, offering welcome shade in the hot desert summer months. From late April until mid-November, it rains down thousands of sweet-smelling magenta blooms. The flowers litter the ground, turning the gravel a lovely shade of purple. I know that blooming begins in late April because I've been tracking this tree regularly since 2011 in *Nature's Notebook*, a plant and animal phenology observing program serving both professional and volunteer scientists. I originally started to observe the phenology of this tree to answer curiosities like whether the onset of flowering varies from one year to the next. After thirteen years of regular observation, I know that flowering can start as early as mid-April and as late as the beginning of May, the number of flowers open on the tree fluctuates over the season, and the seed pods ripen and split open in October and November. Observing this tree has also taught me much more.

Nature's Notebook is one among many "citizen science" phenology programs available worldwide, through which people of all ages and backgrounds can share their observations. Citizen science—also known as community science, volunteer science, participatory science, and public science—engages participants in at least one stage of the process of science, most frequently in data collection, but this can also include stating a question, analyzing the data, or communicating the findings. Much recent discussion has centered on the best language to use to refer to efforts that engage amateur or volunteer scientists in various aspects of scientific research. Several terms have been proposed, though each suffers from shortcomings, and terminology continues to be an actively debated topic within the field. These efforts are widely appreciated for their capacity to generate data and observations at scales far beyond what professional scientists can achieve when working alone. In turn, project participants benefit through what are often novel experiences, and many participants report learning new things and enjoying the social aspects of the work. Thousands of projects from around the world are listed on SciStarter (scistarter.org), a searchable online citizen science hub, engaging participants in activities such as tracking rainfall, water quality sampling, documenting roadkill, and monitoring changing coastlines.

Nature's Notebook (naturesnotebook.org), operated by the USA-NPN, invites participants to track what they see happening in plants by making observations on the same organisms throughout the year. Observations entail tracking not only whether events like flowering are occurring but the intensity or abundance of the event too, such as the proportion of the flower buds on the plant that are open. This additional information more fully characterizes resource availability and animal activity over the course of the season, offering insights into the impacts of events such as late spring freezes, capturing large-scale flowering or fruiting events, and supporting the identification of emerging mismatches between interacting species.[3] The observations I've collected on my desert willow tree reveal that this tree flowers continuously throughout the summer, breaking from the approach utilized by most desert plants of flowering in short bouts following rainfall events. I also track bird, insect, and lizard phenology through *Nature's Notebook*.

Collecting these observations can be a lot of fun. When my boys were younger, I loved to involve them in answering the observation questions for plants in our yard. We would race to see who could find the first young leaf or open flower on our grapefruit and desert willow trees, and adored discovering crab spiders tucked up in the desert willow's trumpet-shaped blooms. A second project, Budburst (budburst.org), invites volunteers to document plant phenology in the United States. Observation protocols are less detailed than those found in *Nature's Notebook*, and as such, Budburst is a better match for younger participants. Both programs use status protocols. This means that each time an observer looks at a plant, they are invited to answer "yes" or "no" to questions such as, "Do you see open flowers?" Given this, both programs help to paint a picture of when phenological events like flowering are occurring and when they are not.

Volunteer phenology observing programs exist in many countries, and include ClimateWatch (climatewatch.org.au) in Australia, SeasonWatch (seasonwatch.in) in India, Observatoire des Saisons (obs-saisons.fr) in France, Natuurkalender (naturetoday.com/intl/nl/observations/natuurkalender) in the Netherlands, Nature's Calendar (naturescalendar.org.uk) in the United Kingdom, the Japan Phenology Monitoring Network in Japan, PhaenoNet (phaenonet.ch/de/) in Switzerland, the Phenological Network of Catalonia (Fenocat, meteo.cat/wpweb/climatologia/fenologia/) in Catalonia, Spain, and Phenoclim (phenoclim.org), which focuses on mountain areas of France, Spain, Italy, and Switzerland. In addition, the Bhutan Phenology Network (heroes.uwicer.gov.bt/home) engages schoolchildren in tracking the phenology of 110 plant species. A wonderful feature of tracking phenology as a volunteer in any of these programs is that it can be done by anyone, anywhere—in your yard, the alley behind your apartment building, the neighborhood park down the street, and even a local cemetery. Both Mount Auburn Cemetery in Boston, Massachusetts, and the Green-Wood Cemetery in Brooklyn, New York, host robust volunteer phenology monitoring programs using *Nature's Notebook*, and participants often remark on the soothing experience of observing plants in these restful spaces.

Several other biodiversity-focused citizen science programs indirectly capture phenology information. Two of the largest are iNaturalist (inaturalist.org) and eBird (ebird.org), both with global reach. The primary focus of both efforts is to document observations of plants and animals (in the case of iNaturalist) and birds (in the case of eBird). The consequent "presence records," which record when and where a species was present, are foundational for determining the habitats where different species are found, how abundant they are, and whether their population numbers are changing. Participants in iNaturalist frequently submit photos of their observations, thereby creating a permanent record of the report that can be consulted for accuracy. The eBird program capitalizes on the long-standing approach among birders to track observations using checklists. Both programs have been wildly successful, with each amassing over a hundred million records as of 2023, which is far, far more than paid scientists and professionals could ever hope to collect.

Tracking phenology can be done by anyone, anywhere—in your yard, the alley behind your apartment building, the neighborhood park down the street, and even in a local cemetery.

Volunteers' Phenology Observations Make Scientific Advancements Possible

As of this writing, over 160 scientific studies have been published that use phenology observations contributed to *Nature's Notebook*. The 40 million records of plant and animal phenology contributed have been used to document recent changes in phenology, establish the cues to leaf out and flowering, ground truth imagery collected by airborne sensors, and build and improve predictive models and forecasts of plant and animal activity. The geographic scale of the observations is a true asset, enabling comparisons across regions with different conditions. For example, a study using observations of leaf out in understory forest shrubs contributed by *Nature's Notebook* participants across the eastern United States showed that invasive shrubs leaf out as much as 77 days earlier than native shrubs, though the duration separating leaf out among the two groups shrinks the further north you go.[4]

Observations contributed to *Nature's Notebook* are also increasingly valuable in management applications. One of the most important factors shaping wildfire risk is plant moisture content—the amount of water in living plants relative to their dry weight. When this ratio, referred to as live fuel moisture, falls below a critical threshold of 79 percent, the size of fires increases dramatically, making it a critically important variable to measure over the course of a fire season. Traditional approaches for estimating live fuel moisture involve clipping and drying vegetation samples, or interpreting imagery collected by drones or satellites. Both approaches suffer from costs and delays between when data are collected and when results are available. Visually assessing the phenological status of individual plants offers a third option for assessing the moisture status in vegetation. A group of researchers recently showed that chamise, a common and widely distributed shrub in fire-prone chaparral ecosystems in California, crosses below the critical live fuel moisture threshold of 79 percent after the plant has flowered but before fruits have developed, using observations contributed to *Nature's Notebook*.[5] With this knowledge, resource managers can easily and inexpensively assess live fuel moisture status in chaparral simply by looking at the flowering and fruiting status of chamise.

Even though the iNaturalist and eBird programs are not squarely focused on documenting phenology, these programs are generating critical phenology information that is enabling important insights into species' phenology and how they are changing. For example, ecologists used reports of yucca plants collected in the southwestern US desert contributed to iNaturalist to determine the typical flowering period for the plants and document anomalous bouts of flowering.[6] Brazilian researchers used iNaturalist participants' observations of frogs and toads to better understand the timing and duration of the animals' reproductive period. The hundreds of photographs of nests, eggs, tadpoles, and frogs with inflated air sacs largely confirmed what was already known about when the critters breed in this region. The broad geographic scope of the iNaturalist reports revealed previously undocumented regional variation in the breeding period, however, and specifically, that frogs located in warmer and drier areas had shorter reproductive periods.[7]

Bird checklists submitted to eBird have similarly enabled novel findings at previously unachievable large geographic scales. In 2012, Drs. Allen Hurlbert and Zhongfei Liang evaluated eBird reports for eighteen species of migratory birds in North America. In the first study of its kind, the scientists demonstrated that species migrating shorter distances exhibited greater advancements in their spring migration arrival dates than long-distance migrants.[8] Since that time, several more studies have used eBird records to document large-scale migration patterns, including different rates of migration among short- and long-distance migrants. More recently, a team of ornithologists determined that American kestrels migrating shorter distances vary their spring arrival time in response to temperatures, while kestrels migrating longer distances do not.[9] These findings match the prevailing narrative that long-distance migrants may be more vulnerable to changes in spring phenology and thus become mismatched from their food availability.

Present-day phenology observations contributed by volunteers in phenology-themed citizen science programs are being combined with historical observations to tell a crucial story of change as well. The Nature's Calendar program in the United Kingdom has engaged volunteers in tracking plant phenology since 2000. When observations contributed to this program were combined with the Marsham family's famous observations from the eighteenth and nineteenth centuries along with observations documented through the Royal Meteorological Society (1875–1947), researchers demonstrated dramatic advancement in spring flowering time for over four hundred plant species in the United Kingdom. Between Marsham's time and the mid-twentieth century, flowering in wood anemone, an early spring flowering perennial plant with attractive yellow and white flowers, has advanced by over three weeks.[10] And since the mid-1980s, when the effects of global warming really started to materialize, spring flowering has advanced by more than a month as temperatures have steadily increased.[11]

A similar chain of events has unfolded in Canada. Between the 1930s to the 1960s, Agriculture Canada tracked phenology in dozens of plants with the aim of supporting agricultural practices. In 1987, the citizen science phenology observing program, PlantWatch, was launched with the aim of better understanding the effects of global warming and climate change. When these datasets were recently combined, researchers found that early spring plants of high latitudes had advanced their flowering times by two weeks since the mid-1900s.[12] As well, in the western European Alps, phenology observations collected by volunteers show clear advancements in bud burst in common birch and European ash.[13]

It's not just scientists and resource managers who care about phenology; the thirst for phenology information and knowledge is popping up in all sorts of surprising places. A few years ago, producers for Mindy Kaling's show *The Sex Lives of College Girls* called the USA-NPN office asking when tree canopies on various college campuses in the northeastern and southeastern United States would be fully leafed out. Though they planned to film in the spring, the show was to be set in the autumn, so they wanted tree canopies full of mature leaves. Using observations of leaf phenology contributed to *Nature's Notebook*, my team at the USA-NPN and I helped the producers determine the location with the greatest chances of fully leafed-out trees when they planned to film. The production crew then recolored the tree canopies in postproduction after consulting with us about appropriate colors for trees in Vermont, the purported location of the fictional campus. We have also been asked when species are in flower by authors wishing to accurately portray facts in their books and even when trees would be fully leafed out by Motorola

Solutions engineers wishing to determine the best time to test attenuation in wireless signal strength.

Reconnecting with Seasonal Changes Is Good for Mind, Body, and Soul

My regular desert willow tree observations harbor similarly useful information. I often reference the observations I logged in previous years to settle disputes with my husband over whether bloom is earlier or later than usual in a particular spring. The records are also helpful for determining how the plants respond to a very wet or very dry monsoon season. My practice of observing this tree and other plants in my yard has given me so much more, though, bringing me peace, connection, grounding, and joy.

Case in point: one summer, we had *many* hawk moth larvae on our desert willow tree. It was great entertainment for my young boys to visit the tree throughout the summer and search for the caterpillars—some of which reached the size of a hot dog! Oddly, the next year, we only had a few. I don't think I've seen more than one or two on the tree in any year since. I have also delighted in my discovery of lynx spiders blending in with green foliage and verdin nests tucked in among the leaves. I would have missed all of these gifts—and the joy of discovery they conferred—had I not been visiting the tree on a regular basis and gazing up into the foliage.

A growing body of evidence supports the idea that maintaining connections with nature is good for our health. A UK-based research team recently quantified the benefits of brief but regular nature-based activities such as surveying butterflies or sitting in nature and recording observations. Participants reported clear improvements in their well-being following the activities, including increased happiness, connectedness with nature, and satisfaction with life.[14] Spending time in nature lowers blood pressure, heart rate, and cortisol levels too. Hospital patients with a view of a green space are released sooner than those with views of only the built environment. Employees with regular contact with nature report fewer sick days and greater job satisfaction. In response, researchers in Australia are now urging their government to incorporate nature-based citizen science into their public health policies to improve public health, especially in urban areas.[15]

After tracking the phenology of plants and animals in my yard for nearly fifteen years, a theory is forming in my head. The repeated act of observing my desert willow tree, year after year, has sparked a sense of anticipation for what might happen next. I am beginning to appreciate the reappearance of seasonal phenomena like flowering in our torch cacti, with their massive and luminous magenta, tangerine, lemon-yellow, and apricot blooms,

A growing body of evidence supports the idea that maintaining connections with nature is good for our health.

like the return of a loved one who's been away for many months. Joy and excitement well within me on the discovery of a flower bud, and I anticipate its opening with glee. And I'm not the only person who gets energized by this. My friend and colleague Jeff, who also hosts a bevy of torch cacti in his yard, and I eagerly share reports every time one of these wonderful plants graces our yards with another bloom. What's so fun is that even though we live miles apart, our plants flower on pretty much the same day every time! I think people stick with their routines of regularly collecting observations of plants and animals through *Nature's Notebook* or other programs in anticipation of what they might see each day.

There's plenty of evidence demonstrating people's enthusiasm for the reappearance of seasonal phenomena. Flowering in night-blooming cereus cacti is a fantastic example. For nearly the entirety of the year, they look like a pile of snakes, and in the case of my yard, a pile of dead snakes. But on one magical night, they open their outrageous blooms, defying what seems possible for a plant you might mistake for yard waste. The blooms, composed of dozens of feathery petals surrounded by a crown of sepals and reaching the size of a volleyball, open for just a few precious hours. What's more, neighboring plants all open their extravagant flowers on the same night. Several botanical gardens in Arizona organize "bloom watch" events as the flowering time approaches, providing frequent flower status updates via websites and emails. The night the buds burst open in all of their glory, thousands of people gather to gaze in appreciation, and the feeling is celebratory. The tens of thousands of people who gather to experience spring and fall bird migrations at Point Pelee National Park in southern Ontario—and so many other locations around the globe—is further evidence that people appreciate and anticipate seasonal events.

Stubbornly persistent activity through truly challenging conditions is further confirmation of how much a regular practice of observing living things matters to people. In spring 2020, when COVID lockdowns were most intense and widespread, iNaturalist and eBird both experienced swift growth in participation.[16] Because access to parks and natural areas where many people collected their phenology observations was cut off, participation in *Nature's Notebook* dropped off in these early months of the pandemic. But by fall 2020, participation had rebounded to pre-COVID levels, as many participants established monitoring sites at new, accessible locations and kept on going.

Observing Can Be a Salve

A few years ago, a beloved member of the USA-NPN passed on. Patty served as our team's botanist, and contributed to observation protocols and guidance offered to *Nature's Notebook* participants. Her close attention to detail, honed through many years as a watercolor artist, was reflected in her work, and the precise and careful definitions of fruits and other botanical terms incorporated into the USA-NPN observation protocols are a small bit of Patty's legacy. Her love of plants helped us all increase our appreciation for their nuance and detail, and her gentle nature brought calm and warmth to the office.

To honor and memorialize all that Patty shared with us, my team and I sponsored a foothills palo verde tree outside our building on the University of Arizona campus in her name. This tree graces us each spring with a spectacular abundance of yellow flowers. The blossoms are diminutive, though on close inspection, they reveal themselves as remarkably beautiful. Five small, triangular, lemon-yellow petals embrace a pistil and several bright orange anthers—the plants' reproductive parts—detail I think Patty would have appreciated. Individual flowers are perched on short, arcing stalks along each branch, forming cascades of flowers that overtake the tree's canopy. Thousands of flowers open synchronously, creating an explosion of color worthy of comment by passing strangers.

Making observations on Patty's tree helps me feel connected to her. The quiet, focused action of observing brings Patty's serene nature to mind. It also reminds me of her appreciation for all things green. By observing this tree, I've noticed that though the major bloom event occurs in the spring, small numbers of flowers appear at other times of the year as well. This is especially true in the later months of the year—October, November, and even into December. Patty would have known that the trees can put on flowers at odd times of year in response to watering. I think she would have known that being present with the tree is good for our souls too.

Patty's tree is situated just outside my office building. By the time I arrive at the tree in the morning, I've already shepherded unwilling boys to school, negotiated Tucson traffic and campus parking, and faced news headlines peppered with climate doom. Pausing for a moment beneath the tree's welcoming branches allows me to catch my breath, physically and metaphorically. Focusing my scattered thoughts on what the tree is doing and how it has changed since my last visit restores my appreciation for the amazing things nature does even when we aren't watching. Sometimes I receive a bonus gift: a butterfly will reveal itself to me or maybe I'll get a glimpse of the Cooper's hawks that have nested in a nearby tree in previous years. This centering activity resets me, restoring a sense of appreciation and gratitude, setting me on a positive course for the next phase of my day.

What Comes Next?

Living in an era of climate change is difficult. Finding ways to stay positive requires us to find ways to reenergize ourselves. For me, that means spending time outside as well as reconnecting with friends and family—including my plant friends. Making connections with other living things brings me joy, meaning, and hope at a personal scale, and this recharges me to work in a field that is being heavily impacted by a rapidly changing climate. My commitment to regularly documenting phenological status on plants in my yard ensures I don't go too long without a dose of this necessary medicine.

It's now midsummer, and mama hummer is at it again, preparing her nest for a second clutch of eggs this season. She inspires me with her determination, carefully lining her nest with soft grasses and wisps of spider webs gathered from the yard. She carries on with this work despite the heat—we have broken records this summer for the number of days over 110°F, and today it topped 112°F here at this space we share with the hummingbirds, lynx spiders, hawk moths, Gambel's quail, desert willow, cacti, and myriad other living things. I am grateful to bear witness to mama hummer's indefatigable efforts to raise young this summer and will note her absence after she migrates south for the winter. I will joyfully anticipate her return to her precarious home base perched on the string of twinkle lights next May. Pulled along by my anticipation of her return, I'll keep a watchful eye throughout the spring. I'll celebrate it with an excited shout to my family members and a record in *Nature's Notebook*.

We protect and care for what we know. National Park Service interpreter Freeman Tilden wrote, "Through interpretation, understanding; through understanding, appreciation; and through appreciation, protection."[17] I'm grateful for the feelings of kinship and intimacy, and the amplified awareness of the seasonal ebb and flow I've gained through observing living things with which I share this space. My wish is that everyone might experience such a connection—or reconnection—with their environment, and find joy, peace, and restoration. We need this mindset to face the changes taking place now and into the future. Though regularly visiting plants and animals in your yard, a nearby park, or abandoned field down the street and documenting what you see might not feel like much, the action can contribute valuable information to support scientific discovery as well as reorient your heart and mind in a way that other efforts cannot. This small step may be the most important one you can take, for yourself and the planet.

ACKNOWLEDGMENTS

I am truly grateful for the opportunity to share my knowledge and enthusiasm for phenology in the form of this book. This book came about after I gave a presentation at a Harvard Radcliffe Institute Science Symposium, where I met MIT Press editor Beth Clevenger. Thanks are due to Immaculata De Vivo and Rebecca Wassarman, who invited me to the event and made this exchange possible. I am thankful to Beth and Anthony Zannino at the MIT Press for their encouragement, patience, and support.

I extend my appreciation to the many individuals who responded to my inquiries and generously shared their knowledge for this manuscript, including Kerissa Fuccillo Battle, Dave Bertelsen, Rob Croll, Melanie Culver, Stanley Fineman, Jennifer Fitchett, Jennifer Gee, Phil Guertin, Samantha Chisholm Hatfield, Dan Herms, Jenny Keroack, Trevor Lantz, Carla Messinger, Melonee Montano, Bill Peachey, Katy Prudic, Jorge Santiago-Blay, and Nancy Turner.

I am also thankful to my family and friends who provided consistent encouragement and enthusiasm. Thank you so much, Mom, Dad, Jo, Dennis, Tina, Andrew, Colleen, Jeff C., Mindy, Joe, Emily, Jeff M., Katy, Jeff O., and Kristin.

Thank you, wonderful members of the USA National Phenology Network team, past and present, especially Nathan Acosta, Samantha Brewer, Ellen Denny, Erin Posthumus, Alyssa Rosemartin, Jeff Switzer, and our cofounders, Julio Betancourt and Mark Schwartz. There are too many wonderful colleagues to mention who have shaped me along my path, but I want to give a shout-out to those friends who listened patiently and cheered me on during this project, especially Caryn Beiter, Amanda Gallinat, and Rob Guralnick.

I am also thankful to Jan Dizard, Joan Maloof, and an anonymous reviewer who provided supportive comments and helpful feedback that significantly improved this book.

Most important, I am forever grateful to my wonderful husband, Mike, and my amazing boys, Zach and Nick, who all tolerated my intense focus on this project throughout summer 2023 and provided (nearly) endless encouragement. Lastly, I must thank Ember, my cat, who consistently and insistently woke me up at 5:00 a.m. to write, and still does, even though the project is now complete.

SPECIES LIST

American kestrel—*Falco sparverius*

American robin—*Turdus migratorius*

arctic caribou—*Rangifer tarandus*

arctic snow goose—*Anser caerulescens*

beluga whale—*Delphinapterus leucas*

bitter root—*Lewisia rediviva*

black bear—*Ursus americanus*

black cherry—*Prunus serotina*

black swallowwort—*Cynanchum louiseae*

black walnut—*Juglans nigra*

black-headed beard runner—*Leistus terminates*

blackpoll warbler—*Setophaga striata*

black-throated blue warbler—*Setophaga caerulescens*

bloodroot—*Sanguinaria canadensis*

blue dicks—*Dichelostemma capitatum*

Brazilian free-tailed bat—*Tadarida brasiliensis*

broad-billed hummingbird—*Cyanthus latirostris*

brown bear—*Ursus artcos*

brown-veined white butterfly—*Benenois aurota*

buffel grass—*Cenchrus ciliaris*

butterfish—*Peprilus triacanthus*

Canada anemone—*Anemone canadensis*

chamise—*Adenostoma fasciculatum*

cheatgrass—*Bromus tectorum*

cherry—*Prunus* spp.

chipping sparrow—*Spizella passerina*

cicada—*Cicada* spp.

cogongrass—*Imperata cylindrica*

common lilac—*Syringa vulgaris*

common milkweed—*Asclepias syriaca*

Cooper's hawk—*Accipiter cooperii*

corn—*Zea mays*

cutleaf toothwort—*Cardamine concatenata*

desert honeysuckle—*Anisacanthus thurberi*

desert tortoise—*Gopherus agassizii*

desert willow—*Chilopsis linearis*

dingy skipper—*Erynnis tages*

dwarf salamander—*Eurycea quadridigitata*

eastern phoebe—*Sayornis phoebe*

emerald ash borer—*Agrilus planipennis*

European adder—*Vipera berus*

Fender's blue butterfly—*Icaricia icarioides fenderi*

field mustard—*Brassica rapa*

flowering dogwood—*Cornus florida*

foothills palo verde—*Parkinsonia microphylla*

forsythia—*Forsythia* spp.

Gambel's quail—*Callipepla gambelii*

grasshopper sparrow—*Ammodramus savannarum*

grass-leaf aphid—*Utamphorophora humboldti*

great crested flycatcher—*Myiarchus crinitus*

green-veined white butterfly—*Pieris napi*

hawk moth—*Sphingidae* spp.

hepatica—*Anemone hepatica*

horse chestnut—*Aesculus hippocastanum*

house wren—*Troglodytes aedon*

Indian jujube—*Ziziphus mauritiana*

Indian laburnum—*Cassia fistula*

Indian pipes—*Monotropa uniflora*

indigo bunting—*Passerina cyanea*

ironwood—*Ostrya virginiana*

jacaranda—*Jacaranda mimosifolia*

Japanese barberry—*Berberis thunbergii*

least flycatcher—*Empidonax minimus*

longfin squid—*Doryteuthis pealeii*

longhorn sculpin—*Myoxocephalus octodecemspinosus*

lynx spider—*Peucetia viridans*

maple—*Acer* spp.

mayfly—*Ephemeroptera* spp.

Mexican long-tongued bat—*Choeronycteris Mexicana*

monarch—*Danaus plexippus*

Namaqualand daisy—*Dimorphotheca aurantiaca*

nannyberry—*Viburnum lentago*

narwhal—*Monodon monoceros*

night-blooming cereus—*Cereus greggii*

North American deer mice—*Peromyscus maniculatus*

Ohio buckeye—*Aesculus glabra*

ornate chorus frog—*Pseudacris ornata*

Parry's false prairie clover—*Marina parryi*

Pennsylvania smartweed—*Polygonum pensylvanicum*

pine warbler—*Setophaga pinus*

plum tree—*Prunus domestica*

red brome—*Bromus rubens*

red chokecherry—*Prunus virginiana*

red columbine—*Aquilegia canadensis*

red maple—*Acer rubrum*

red-winged blackbird—*Agelaius phoeniceus*

reed warbler—*Acrocephalus scirpaceus*

Richardson's squirrel—*Urocitellus richardsonii*

rock grayling butterfly—*Hipparchia semele*

rosebay—*Rhododendron maximum*

saffron—*Crocus sativus*

saguaro—*Carnegiea gigantea*

Saint-John's wort—*Hypericum perforatum*

salmonberry—*Rubus spectabilis*

sandwich tern—*Thalasseus sandvicensis*

Scarlet tanager—*Piranga olivacea*

scup—*Stenotomus chrysops*

shad—*Alosa sapidissima*

shadbush—*Amelanchier canadensis*

sixweeks threeawn—*Aristida adscensionis*

skunk cabbage—*Symplocarpus foetidus*

smooth sumac—*Rhus glabra*

snowshoe hare—*Lepus americanus*

spring peeper—*Pseudacris crucifer*

striped searobin—*Prionotus evolans*

sugar maple—*Acer saccharum*

summer flounder—*Paralichthys dentatus*

Swainson's thrush / salmonberry bird—*Catharus ustulatus*

tamarind—*Tamarindus indicus*

Tatarian honeysuckle—*Lonicera tatarica*

tiger salamander—*Ambystoma tigrinum*

tulip tree—*Liriodendron tulipifera*

Vanhoutte spirea—*Spiraea* × *vanhouttei*

violet ground beetle—*Carabus violaceus*

water flea—*Bythrotrephes longimanus*

waxwing—*Bombycilla* ssp.

white-crowned sparrow—*Zonotrichia leucophrys*

white dogwood—*Cornus florida*

white-winged dove—*Zenaida asiatica*

wild blue phlox—*Phlox divaricata*

winter sardine—*Sardina pilchardus*

wood anemone—*Anemone numerosa*

wood duck—*Aix sponsa*

wood warbler—*Phylloscopus sibilatrix*

yellow bells—*Tecoma stans*

yellow sweet clover—*Melilotus officinalis*

yellow trout lily—*Erythronium americanum*

yellow wood sorrel—*Oxalis stricta*

GLOSSARY

Brumation

Brumation is a hibernation-like state that some reptiles and amphibians enter during the colder months of the year. This physiological response is an adaptation to survive unfavorable environmental conditions, such as low temperatures and reduced food availability.

Citizen science

Citizen science, also known as community science or public participation in scientific research, is a collaborative approach to scientific inquiry in which members of the public who may not have formal scientific training actively participate in various aspects of scientific research. In citizen science projects, volunteers, often referred to as citizen scientists, contribute their time, skills, and observations to collect, analyze, or interpret data, and sometimes even help formulate research questions.

Correlation

Correlation reflects the degree of relationship between two or more variables. In the field of phenology, correlation is often used to assess whether a phenological transition, such as leaf out or flowering, is associated with particular environmental conditions. This information can be used to make predictions about when phenological events might occur.

Cue

Environmental cues are observable stimuli that trigger a biological response or behavioral change in organisms. Common cues include daylength, temperature, and moisture conditions.

Diapause

Diapause is a biological phenomenon characterized by a temporary pause or suspension of metabolic and developmental activities in an organism, typically in response to unfavorable environmental conditions. It is a survival strategy employed by various organisms, including insects, mammals, and plants, to endure harsh conditions such as extreme temperatures, drought, or food scarcity.

Dormancy

Dormancy is a state of reduced physical and metabolic activity that some organisms enter into as a response to adverse environmental conditions or specific triggers. During dormancy, an organism temporarily suspends or slows down various life processes, such as growth, reproduction, and mobility to conserve energy and survive unfavorable circumstances.

Leaf senescence

Leaf senescence is the natural aging process that occurs in plant leaves as they approach the end of their functional life span. During leaf senescence, leaves undergo a series of physiological, biochemical, and structural changes that ultimately lead to their deterioration and eventual shedding from the plant.

Mathematical model

A mathematical model is a simplified representation of a real-world system or phenomenon using mathematical equations, symbols, and relationships. These models are designed to capture essential characteristics and behaviors of the system being studied, allowing scientists and researchers to analyze, predict, and understand the phenomenon's behavior under various conditions.

Migration

Migration refers to the seasonal or periodic movement of individuals or groups of organisms from one geographic location to another. This movement is often driven by factors such as changes in environmental conditions, availability of resources, or the need to complete specific life processes.

Phenological mismatch

Phenological mismatch refers to a temporal misalignment in the timing of life cycle events between interacting species. This phenomenon occurs when one species, typically dependent on another for resources or interactions, undergoes a critical biological event (such as flowering, nesting, or hatching) either earlier or later in relation to the timing of an interacting species.

Phenology

Phenology is the branch of ecology and biology that focuses on the study of recurring biological events and life cycle stages in various organisms, particularly plants and animals, in relation to seasonal and environmental changes.

Phenophase
An observable stage or phase in the annual life cycle of a plant or animal that can be defined by a start and end point. Phenophases generally have a duration of a few days or weeks. Examples include the period over which newly emerging leaves are visible or open flowers are present on a plant.

Remote sensing
Remote sensing is a technique used in earth and environmental sciences to gather information about the earth's surface and atmosphere from a distance, typically using sensors or instruments aboard satellites, aircraft, drones, or ground-based systems. It involves the collection of data, such as imagery, spectral information, and other measurements, without direct physical contact with the target area or object being observed.

Seasonality
Seasonality refers to the regular and predictable patterns of change or variation that occur in nonliving phenomena over the course of a year. These patterns are often associated with the earth's axial tilt and its orbit around the sun, leading to distinct changes in weather, temperature, and daylength during different times of the year.

NOTES

Chapter 1

1. Julianna J. Renzi, William D. Peachey, and Katharine L. Gerst, "A Decade of Flowering Phenology of the Keystone Saguaro Cactus (*Carnegiea gigantea*)." *American Journal of Botany* 106, no. 2 (2019): 199–210.
2. Helmut Lieth, "Purposes of a Phenology Book," in *Phenology and Seasonality Modeling*, ed. Helmut Lieth (New York: Springer-Verlag, 1974), 3–20.
3. Gaston R. Demarée and This Rutishauser, "From 'Periodical Observations' to 'Anthochronology' and 'Phenology': The Scientific Debate between Adolphe Quetelet and Charles Morren on the Origin of the Word 'Phenology,'" *International Journal of Biometeorology* 55 (2011): 753–761.
4. Nancy Turner, *Ancient Pathways, Ancestral Knowledge: Ethnobotany and Ecological Wisdom of Indigenous Peoples of Northwestern North America*, vol. 74 (Montreal: McGill-Queen's University Press, 2014).
5. Thomas Richmond Kenote, "Indigenous Phenology: An Interdisciplinary Case Study on Indigenous Phenological Knowledge on the Menominee Nation Forest" (PhD diss., University of Minnesota, 2020).
6. Trevor C. Lantz and Nancy J. Turner, "Traditional Phenological Knowledge of Aboriginal Peoples in British Columbia," *Journal of Ethnobiology* 23, no. 2 (2003): 263–286.
7. Carla Messinger and Susan Katz, *When the Shadbush Blooms* (New York: Lee and Low Books Inc., 2007).
8. C. Chu, "New Monthly Calendar," *Bulletin of the Chinese Meteorological Society* 6 (1931): 1–14; Xiaoqiu Chen, "East Asia," in *Phenology: An Integrative Environmental Science*, ed. Mark D. Schwartz (Dordrecht, South Holland: Springer, 2013), 9–22; Masatoshi Yoshino, "Development of Phenological Recognition and Phenology in Ancient China," *Japanese Journal of Biometeorology* 41 (2004): 141–154.
9. Yoshino, "Development of Phenological Recognition," 141.
10. Demarée and Rutishauser, "From 'Periodical Observations' to 'Anthochronology' and 'Phenology,'" 754.
11. Frank N. Egerton, "A History of the Ecological Sciences, Part 23: Linnaeus and the Economy of Nature," *Bulletin of the Ecological Society of America* 88, no. 1 (2007): 79.

12. René A. F. de Réaumur, "Observation du thermometer, faites à Paris pendant l'année 1735, compares avec celles qui ont été faites sous la ligne, à l'Isle de France, à Alger et en quelques-unes de nos isles de l'Amérique," *Mémoires de l'Académie des Sciences de Paris (1735)*.
13. Gustav Jahoda, "Quetelet and the Emergence of the Behavioral Sciences," *Springerplus* (2015): 473.
14. Andrew D. Hopkins, "The Bioclimatic Law as Applied to Entomological Research and Farm Practice," *Scientific Monthly* 8 (1919): 496–513.
15. Martin L. Parry, Osvaldo Canziani, Jean Palutikof, Paul Van der Linden, and Clair Hanson, eds., *Climate Change 2007: Impacts, Adaptation, and Vulnerability: Working Group II Contribution to the Fourth Assessment Report of the IPCC*, vol. 4 (Cambridge: Cambridge University Press, 2007).

Chapter 2

1. "First Warm Day of the Year Red Carpet Cold Open," *Saturday Night Live*, featuring Heidi Gardner and Bowen Yang, aired April 15, 2023, on NBC, https://www.youtube.com/watch?v=UZhjIyGmrRQ.
2. H.-O. Pörtner, D. C. Roberts, E. S. Poloczanska, K. Mintenbeck, M. Tignor, A. Alegría, M. Craig, et al., "Summary for Policymakers," in *Climate Change 2022: Impacts, Adaptation, and Vulnerability: Contribution of Working Group II to the Sixth Assessment Report of the Intergovernmental Panel on Climate Change*, eds. H.-O. Pörtner, D. C. Roberts, M. Tignor, E. S. Poloczanska, K. Mintenbeck, A. Alegría, M. Craig, et al. (Cambridge: Cambridge University Press, 2022), 3–33, doi:10.1017/9781009325844.001.
3. "Past Eight Years Confirmed to Be the Eight Warmest on Record," World Meteorological Organization, January 12, 2023, https://wmo.int/media/news/past-eight-years-confirmed-be-eight-warmest-record.
4. R. Iestyn Woolway, Benjamin M. Kraemer, John D. Lenters, Christopher J. Merchant, Catherine M. O'Reilly, and Sapna Sharma, "Global Lake Responses to Climate Change," *Nature Reviews Earth and Environment* 1, no. 8 (2020): 388–403.
5. Paola Arias, Nicolas Bellouin, Erika Coppola, Richard Jones, Gerhard Krinner, Jochem Marotzke, Vaishali Naik, et al., "Climate Change 2021: The Physical Science Basis: Contribution of Working Group I to the Sixth Assessment Report of the Intergovernmental Panel on Climate Change; Technical Summary," in *Climate Change 2021: The Physical Science Basis: Contribution of Working Group I to the Sixth Assessment Report of the Intergovernmental Panel on Climate Change*, ed. V. Masson-Delmotte, P. Zhai, A. Pirani, S. L. Connors,

C. Péan, S. Berger, N. Caud, et al. (Cambridge: Cambridge University Press, 2021), 33–144.
6. Heather M. Kharouba, Johan Ehrlén, Andrew Gelman, Kjell Bolmgren, Jenica M. Allen, Steve E. Travers, and Elizabeth M. Wolkovich, "Global Shifts in the Phenological Synchrony of Species Interactions over Recent Decades," *Proceedings of the National Academy of Sciences* 115, no. 20 (2018): 5211–5216.
7. Stephen J. Thackeray, Peter A. Henrys, Deborah Hemming, James R. Bell, Marc S. Botham, Sarah Burthe, Pierre Helaouet, et al., "Phenological Sensitivity to Climate across Taxa and Trophic Levels," *Nature* 535, no. 7611 (2016): 241–245; Tomas Roslin, Laura Antao, Maria Hällfors, Evgeniy Meyke, Coong Lo, Gleb Tikhonov, Maria del Mar Delgado, et al., "Phenological Shifts of Abiotic Events, Producers and Consumers across a Continent," *Nature Climate Change* 11, no. 3 (2021): 241–248.
8. Michael Stemkovski, James R. Bell, Elizabeth R. Ellwood, Brian D. Inouye, Hiromi Kobori, Sang Don Lee, Trevor Lloyd-Evans, et al., "Disorder or a New Order: How Climate Change Affects Phenological Variability," *Ecology* 104, no. 1 (2023): e3846.
9. Shilong Piao, Qiang Liu, Anping Chen, Ivan A. Janssens, Yongshuo Fu, Junhu Dai, Lingli Liu, et al., "Plant Phenology and Global Climate Change: Current Progresses and Challenges," *Global Change Biology* 25, no. 6 (2019): 1922–1940.
10. Liming Zhou, Compton J. Tucker, Robert K. Kaufmann, Daniel Slayback, Nikolay V. Shabanov, and Ranga B. Myneni, "Variations in Northern Vegetation Activity Inferred from Satellite Data of Vegetation Index during 1981 to 1999," *Journal of Geophysical Research: Atmospheres* 106, no. D17 (2001): 20069–20083; Kirsten M. De Beurs and Geoffrey M. Henebry, "Land Surface Phenology and Temperature Variation in the International Geosphere-Biosphere Program High-Latitude Transects," *Global Change Biology* 11, no. 5 (2005): 779–790; Piao et al., "Plant Phenology and Global Climate Change."
11. Nicole E. Rafferty, Jeffrey M. Diez, and C. David Bertelsen, "Changing Climate Drives Divergent and Nonlinear Shifts in Flowering Phenology across Elevations," *Current Biology* 30, no. 3 (2020): 432–441.
12. Kerissa Fuccillo Battle, Anna Duhon, Conrad R. Vispo, Theresa M. Crimmins, Todd N. Rosenstiel, Lilas L. Armstrong-Davies, and Catherine E. de Rivera, "Citizen Science across Two Centuries Reveals Phenological Change among Plant Species and Functional Groups in the Northeastern US," *Journal of Ecology* 110, no. 8 (2022): 1757–1774.
13. Lori Petrauski, Sheldon F. Owen, George D. Constantz, and James T. Anderson, "Changes in Flowering Phenology of *Cardamine concatenata* and

Erythronium americanum over 111 Years in the Central Appalachians," *Plant Ecology* 220, no. 9 (2019): 817–828.
14. Benjamin I. Cook, Edward R. Cook, Paul C. Huth, John E. Thompson, Anna Forster, and Daniel Smiley, "A Cross-Taxa Phenological Dataset from Mohonk Lake, NY and Its Relationship to Climate," *International Journal of Climatology: A Journal of the Royal Meteorological Society* 28, no. 10 (2008): 1369–1383.
15. Abraham J. Miller-Rushing and Richard B. Primack, "Global Warming and Flowering Times in Thoreau's Concord: A Community Perspective," *Ecology* 89, no. 2 (2008): 332–341.
16. Yann Vitasse, Frederik Baumgarten, C. M. Zohner, T. Rutishauser, B. Pietragalla, R. Gehrig, J. Dai, et al., "The Great Acceleration of Plant Phenological Shifts," *Nature Climate Change* 12, no. 4 (2022): 300–302.
17. Kelsey L. Dunnell and Steven E. Travers, "Shifts in the Flowering Phenology of the Northern Great Plains: Patterns over 100 Years," *American Journal of Botany* 98, no. 6 (2011): 935–945.
18. Marie R. Keatley, Lynda E. Chambers, and Rebecca Phillips, "Australia and New Zealand," *Phenology: An Integrative Environmental Science* (2013): 23–52.
19. Andrea Romano, László Zsolt Garamszegi, Diego Rubolini, and Roberto Ambrosini, "Temporal Shifts in Avian Phenology across the Circannual Cycle in a Rapidly Changing Climate: A Global Meta-Analysis," *Ecological Monographs* 93, no. 1 (2023): e1552.
20. Frank A. La Sorte, Wesley M. Hochachka, Andrew Farnsworth, André A. Dhondt, and Daniel Sheldon, "The Implications of Mid-Latitude Climate Extremes for North American Migratory Bird Populations," *Ecosphere* 7, no. 3 (2016): e01261.
21. Eric K. Waller, Theresa M. Crimmins, Jessica J. Walker, Erin E. Posthumus, and Jake F. Weltzin, "Differential Changes in the Onset of Spring across US National Wildlife Refuges and North American Migratory Bird Flyways," *PLOS One* 13, no. 9 (2018): e0202495; Theresa M. Crimmins and Michael A. Crimmins, "Biologically-Relevant Trends in Springtime Temperatures across the United States," *Geophysical Research Letters* 46, no. 21 (2019): 12377–12387.
22. Casey Youngflesh, Jacob Socolar, Bruna R. Amaral, Ali Arab, Robert P. Guralnick, Allen H. Hurlbert, Raphael LaFrance, et al., "Migratory Strategy Drives Species-Level Variation in Bird Sensitivity to Vegetation Green-Up," *Nature Ecology and Evolution* 5, no. 7 (2021): 987–994; Stephen J. Mayor, Robert P. Guralnick, Morgan W. Tingley, Javier Otegui, John C. Withey, Sarah C. Elmendorf, Margaret E. Andrew, et al., "Increasing Phenological Asynchrony between Spring Green-Up and Arrival of Migratory Birds," *Scientific Reports* 7, no. 1 (2017): 1902.

23. John M. Bates, Mason Fidino, Laurel Nowak-Boyd, Bill M. Strausberger, Kenneth A. Schmidt, and Christopher J. Whelan, "Climate Change Affects Bird Nesting Phenology: Comparing Contemporary Field and Historical Museum Nesting Records," *Journal of Animal Ecology* 92, no. 2 (2023): 263–272; Jacob B. Socolar, Peter N. Epanchin, Steven R. Beissinger, and Morgan W. Tingley, "Phenological Shifts Conserve Thermal Niches in North American Birds and Reshape Expectations for Climate-Driven Range Shifts," *Proceedings of the National Academy of Sciences* 114, no. 49 (2017): 12976–12981.

24. Lucyna Halupka, Andrzej Dyrcz, and Marta Borowiec, "Climate Change Affects Breeding of Reed Warblers *Acrocephalus scirpaceus*," *Journal of Avian Biology* 39, no. 1 (2008): 95–100; Andrea K. Townsend, T. Scott Sillett, Nina K. Lany, Sara A. Kaiser, Nicholas L. Rodenhouse, Michael S. Webster, and Richard T. Holmes, "Warm Springs, Early Lay Dates, and Double Brooding in a North American Migratory Songbird, the Black-Throated Blue Warbler," *PLOS One* 8, no. 4 (2013): e59467.

25. Phillip M. Stepanian and Charlotte E. Wainwright, "Ongoing Changes in Migration Phenology and Winter Residency at Bracken Bat Cave," *Global Change Biology* 24, no. 7 (2018): 3266–3275.

26. Heather E. Johnson, David L. Lewis, Tana L. Verzuh, Cody F. Wallace, Rebecca M. Much, Lyle K. Willmarth, and Stewart W. Breck, "Human Development and Climate Affect Hibernation in a Large Carnivore with Implications for Human–Carnivore Conflicts," *Journal of Applied Ecology* 55, no. 2 (2018): 663–672; M. M. Delgado, G. Tikhonov, E. Meyke, M. Babushkin, T. Bespalova, S. Bondarchuk, A. Esengeldenova, et al., "The Seasonal Sensitivity of Brown Bear Denning Phenology in Response to Climatic Variability," *Frontiers in Zoology* 15 (2018): 1–11.

27. Natacha Larnaud, "Climate Change Is Causing Bears to Come Out of Hibernation a Month Early—Which Can Be Dangerous for Humans," CBS News, March 12, 2020, https://www.cbsnews.com/news/climate-change-bears-come-out-of-hibernation-month-early.

28. Valerie Bourdeau, "Climate Change Is Cramping These Squirrels' Love Life," Weather Network, April 15, 2023, https://www.theweathernetwork.com/en/news/climate/impacts/climate-change-cramping-richardson-ground-squirrels-love-life.

29. Brian D. Todd, David E. Scott, Joseph H. K. Pechmann, and J. Whitfield Gibbons, "Climate Change Correlates with Rapid Delays and Advancements in Reproductive Timing in an Amphibian Community," *Proceedings of the Royal Society B: Biological Sciences* 278, no. 1715 (2011): 2191–2197.

30. Rebecca K. Turner and Ilya M. D. Maclean, "Microclimate-Driven Trends in Spring-Emergence Phenology in a Temperate Reptile (*Vipera berus*): Evidence for a Potential 'Climate Trap'?," *Ecology and Evolution* 12, no. 2 (2022): e8623.
31. Bengt Karlsson, "Extended Season for Northern Butterflies," *International Journal of Biometeorology* 58 (2014): 691–701.
32. James R. Bell, Lynda Alderson, Daniela Izera, Tracey Kruger, Sue Parker, Jon Pickup, Chris R. Shortall, et al., "Long-Term Phenological Trends, Species Accumulation Rates, Aphid Traits and Climate: Five Decades of Change in Migrating Aphids," *Journal of Animal Ecology* 84, no. 1 (2015): 21–34.
33. Gabor Pozsgai and Nick A. Littlewood, "Ground Beetle (Coleoptera: Carabidae) Population Declines and Phenological Changes: Is There a Connection?," *Ecological Indicators* 41 (2014): 15–24.
34. Karlsson, "Extended Season for Northern Butterflies," 694.
35. John Lyons, Andrew L. Rypel, Paul W. Rasmussen, Thomas E. Burzynski, Bradley T. Eggold, Jared T. Myers, Tammie J. Paoli, and Peter B. McIntyre, "Trends in the Reproductive Phenology of Two Great Lakes Fishes," *Transactions of the American Fisheries Society* 144, no. 6 (2015): 1263–1274.
36. Marina M. Manca, Magda Portogallo, and Meghan E. Brown, "Shifts in Phenology of Bythotrephes Longimanus and Its Modern Success in Lake Maggiore as a Result of Changes in Climate and Trophy," *Journal of Plankton Research* 29, no. 6 (2007): 515–525.
37. Harvey J. Walsh, David E. Richardson, Katrin E. Marancik, and Jonathan A. Hare, "Long-Term Changes in the Distributions of Larval and Adult Fish in the Northeast US Shelf Ecosystem," *PLOS One* 10, no. 9 (2015): e0137382.
38. Joseph A. Langan, Gavino Puggioni, Candace A. Oviatt, M. Elisabeth Henderson, and Jeremy S. Collie, "Climate Alters the Migration Phenology of Coastal Marine Species," *Marine Ecology Progress Series* 660 (2021): 1–18.
39. Donna D. W. Hauser, Kristin L. Laidre, Kathleen M. Stafford, Harry L. Stern, Robert S. Suydam, and Pierre R. Richard, "Decadal Shifts in Autumn Migration Timing by Pacific Arctic Beluga Whales Are Related to Delayed Annual Sea Ice Formation," *Global Change Biology* 23, no. 6 (2017): 2206–2217.
40. Courtney R. Shuert, Marianne Marcoux, Nigel E. Hussey, Mads Peter Heide-Jørgensen, Rune Dietz, and Marie Auger-Méthé, "Decadal Migration Phenology of a Long-Lived Arctic Icon Keeps Pace with Climate Change," *Proceedings of the National Academy of Sciences* 119, no. 45 (2022): e2121092119.
41. Jonas Jourdan, Viktor Baranov, Rüdiger Wagner, Martin Plath, and Peter Haase, "Elevated Temperatures Translate into Reduced Dispersal Abilities in a

Natural Population of an Aquatic Insect," *Journal of Animal Ecology* 88, no. 10 (2019): 1498–1509.
42. Charles G. Willis, Brad Ruhfel, Richard B. Primack, Abraham J. Miller-Rushing, and Charles C. Davis, "Phylogenetic Patterns of Species Loss in Thoreau's Woods Are Driven by Climate Change," *Proceedings of the National Academy of Sciences* 105, no. 44 (2008): 17029–17033.
43. Rachael E. Bonoan, Elizabeth E. Crone, Collin B. Edwards, and Cheryl B. Schultz, "Changes in Phenology and Abundance of an At-Risk Butterfly," *Journal of Insect Conservation* 25, no. 3 (2021): 499–510.
44. Maria H. Hällfors, Juha Pöyry, Janne Heliölä, Ilmari Kohonen, Mikko Kuussaari, Reima Leinonen, Reto Schmucki, et al., "Combining Range and Phenology Shifts Offers a Winning Strategy for Boreal Lepidoptera," *Ecology Letters* 24, no. 8 (2021): 1619–1632.
45. Lin Meng, Yuyu Zhou, Miguel O. Román, Eleanor C. Stokes, Zhuosen Wang, Ghassem R. Asrar, Jiafu Mao, et al., "Artificial Light at Night: An Underappreciated Effect on Phenology of Deciduous Woody Plants," *PNAS Nexus* 1, no. 2 (2022): pgac046.
46. Daisy Englert Duursma, Rachael V. Gallagher, and Simon C. Griffith, "Effects of El Niño Southern Oscillation on Avian Breeding Phenology," *Diversity and Distributions* 24, no. 8 (2018): 1061–1071.
47. Matthew P. Dannenberg, Conghe Song, Taehee Hwang, and Erika K. Wise, "Empirical Evidence of El Niño–Southern Oscillation Influence on Land Surface Phenology and Productivity in the Western United States," *Remote Sensing of Environment* 159 (2015): 167–180.
48. Theresa M. Crimmins, Michael A. Crimmins, and C. David Bertelsen, "Complex Responses to Climate Drivers in Onset of Spring Flowering across a Semi-Arid Elevation Gradient," *Journal of Ecology* 98, no. 5 (2010): 1042–1051.
49. Rafferty, Diez, and Bertelsen, "Changing Climate Drives Divergent and Nonlinear Shifts."
50. Amy M. Iler, Toke T. Høye, David W. Inouye, and Niels M. Schmidt, "Nonlinear Flowering Responses to Climate: Are Species Approaching Their Limits of Phenological Change?," *Philosophical Transactions of the Royal Society B: Biological Sciences* 368, no. 1624 (2013): 20120489; Sarah C. Elmendorf and Robert D. Hollister, "Limits on Phenological Response to High Temperature in the Arctic," *Scientific Reports* 13, no. 1 (2023): 208.
51. Marcel E. Visser, "Phenology: Climate Change Is Shifting the Rhythm of Nature," in *Frontiers 2022: Noise, Blazes, and Mismatches: Emerging Issues of Environmental Concern*, ed. Pinya Sarasas (Nairobi: United Nations Environment Programme, 2022), 41–58.

Chapter 3

1. Tegan Armarego-Marriott, "The Citizens Who Chart Changing Climate," *Nature Climate Change* 12, no. 4 (2022): 311–312.

2. Nicole E. Rafferty, Jeffrey M. Diez, and C. David Bertelsen, "Changing Climate Drives Divergent and Nonlinear Shifts in Flowering Phenology across Elevations," *Current Biology* 30, no. 3 (2020): 432–441.

3. Theresa M. Crimmins, Michael A. Crimmins, and C. David Bertelsen, "Complex Responses to Climate Drivers in Onset of Spring Flowering across a Semi-Arid Elevation Gradient," *Journal of Ecology* 98, no. 5 (2010): 1042–1051.

4. Anna Ledneva, Abraham J. Miller-Rushing, Richard B. Primack, and Carolyn Imbres, "Climate Change as Reflected in a Naturalist's Diary, Middleborough, Massachusetts," *Wilson Bulletin* 116, no. 3 (2004): 224–231.

5. Caitlin McDonough MacKenzie, Jason Johnston, Abraham J. Miller-Rushing, William Sheehan, Robert Pinette, and Richard Primack, "Advancing Leaf-Out and Flowering Phenology Is Not Matched by Migratory Bird Arrivals Recorded in Hunting Guide's Journal in Aroostook County, Maine," *Northeastern Naturalist* 26, no. 3 (2019): 561–579.

6. Kellen Calinger and Peter Curtis, "A Century of Climate Warming Results in Growing Season Extension: Delayed Autumn Leaf Phenology in North Central North America," *PLOS One* 18, no. 3 (2023): e0282635.

7. Daniel R. Cayan, Susan A. Kammerdiener, Michael D. Dettinger, Joseph M. Caprio, and David H. Peterson, "Changes in the Onset of Spring in the Western United States," *Bulletin of the American Meteorological Society* 82, no. 3 (2001): 399–416; Mark D. Schwartz and Bernhard E. Reiter, "Changes in North American Spring," *International Journal of Climatology* 20, no. 8 (2000): 929–932.

8. Michael A. Crimmins and Theresa M. Crimmins, "Monitoring Plant Phenology Using Digital Repeat Photography," *Environmental Management* 41 (2008): 949–958.

9. Andrew D. Richardson, "PhenoCam: An Evolving, Open-Source Tool to Study the Temporal and Spatial Variability of Ecosystem-Scale Phenology," *Agricultural and Forest Meteorology* 342 (2023): 109751.

10. Matthew L. Forister and Arthur M. Shapiro, "Climatic Trends and Advancing Spring Flight of Butterflies in Lowland California," *Global Change Biology* 9, no. 7 (2003): 1130–1135; Matthew L. Forister, Christopher A. Halsch, C. C. Nice, James A. Fordyce, Thomas E. Dilts, Jeffrey C. Oliver, Kathleen L. Prudic, et al., "Fewer Butterflies Seen by Community Scientists across the Warming and Drying Landscapes of the American West," *Science* 371, no. 6533 (2021): 1042–1045.

11. Samantha M. Knight, Grace M. Pitman, D. T. Tyler Flockhart, and D. Ryan Norris, "Radio-Tracking Reveals How Wind and Temperature Influence the Pace of Daytime Insect Migration," *Biology Letters* 15, no. 7 (2019): 20190327.
12. M. Yu Belyaev, O. N. Volkov, O. N. Solomina, and G. M. Tertitsky, "Animal Migration Studies with the Use of ICARUS Scientific Equipment in the URAGAN Space Experiment aboard the Russian Segment of the ISS," *Gyroscopy and Navigation* 13, no. 3 (2022): 129–140.
13. Marina D. A. Scarpelli, Paul Roe, David Tucker, and Susan Fuller, "Soundscape Phenology: The Effect of Environmental and Climatic Factors on Birds and Insects in a Subtropical Woodland," *Science of the Total Environment* 878 (2023): 163080.
14. Kyle G. Horton, Sara R. Morris, Benjamin M. Van Doren, and Kristen M. Covino, "Six Decades of North American Bird Banding Records Reveal Plasticity in Migration Phenology," *Journal of Animal Ecology* 92, no. 3 (2023): 738–750.
15. Yuting Deng, Maria Carolina T. D. Belotti, Wenlong Zhao, Zezhou Cheng, Gustavo Perez, Elske Tielens, Victoria F. Simons, et al. "Quantifying Long-Term Phenological Patterns of Aerial Insectivores Roosting in the Great Lakes Region Using Weather Surveillance Radar," *Global Change Biology* 29, no. 5 (2023): 1407–1419.
16. Gao Hu, Ka S. Lim, Nir Horvitz, Suzanne J. Clark, Don R. Reynolds, Nir Sapir, and Jason W. Chapman, "Mass Seasonal Bioflows of High-Flying Insect Migrants," *Science* 354, no. 6319 (2016): 1584–1587.
17. Bryan S. McLean and Robert P. Guralnick, "Digital Biodiversity Data Sets Reveal Breeding Phenology and Its Drivers in a Widespread North American Mammal," *Ecology* 102, no. 3 (2021): e03258.
18. Nicholas N. Dorian, Max W. McCarthy, and Elizabeth E. Crone, "Ecological Traits Explain Long-Term Phenological Trends in Solitary Bees," *Journal of Animal Ecology* 92, no. 2 (2023): 285–296.
19. Jennifer M. Fitchett, Antonia Pandazis, and Subhashinidevi Pillay, "Advance in the Timing of the Annual Migration of the Brown-Veined White Butterfly through Johannesburg, South Africa, over the Period 1914–2020," *International Journal of Biometeorology* (2022): 1–8.
20. Jennifer M. Fitchett and Kestrel Raik, "Phenological Advance of Blossoming over the Past Century in One of the World's Largest Urban Forests, Gauteng City-Region, South Africa," *Urban Forestry and Urban Greening* 63 (2021): 127238.
21. Pascal L. Snyman and Jennifer M. Fitchett, "Phenological Advance in the South African Namaqualand Daisy First and Peak Bloom: 1935–2018,"

International Journal of Biometeorology 66, no. 4 (2022): 699–717; Jennifer M. Fitchett, Stefan W. Grab, and Heinrich Portwig, "Progressive Delays in the Timing of Sardine Migration in the Southwest Indian Ocean," *South African Journal of Science* 115, no. 7–8 (2019): 1–6.
22. Dinah Voyles Pulver, "Festivals Forced to Adapt as Climate Change Disrupts Historic Weather Patterns," *USA Today*, March 23, 2022, https://www.usatoday.com/story/news/2022/03/23/climate-change-impact-festival-schedule-challenges/7096240001/?gnt-cfr=1.
23. Pieter De Frenne, Lisa Van Langenhove, Alain Van Driessche, Cédric Bertrand, Kris Verheyen, and Pieter Vangansbeke, "Using Archived Television Video Footage to Quantify Phenology Responses to Climate Change," *Methods in Ecology and Evolution* 9, no. 8 (2018): 1874–1882.
24. A. K. Menzies, Ella Bowles, M. Gallant, H. Patterson, Cory Leigh Kozmik, S. Chiblow, Deborah McGregor, et al., "'I See My Culture Starting to Disappear': Anishinaabe Perspectives on the Socioecological Impacts of Climate Change and Future Research Needs," *Facets* 7, no. 1 (2022): 509–527.
25. Nelson Chanza and Walter Musakwa, "Indigenous Local Observations and Experiences Can Give Useful Indicators of Climate Change in Data-Deficient Regions," *Journal of Environmental Studies and Sciences* 12, no. 3 (2022): 534–546.
26. Menzies et al., "I See My Culture Starting to Disappear."
27. Victoria Reyes-García, Santiago Álvarez-Fernández, Petra Benyei, David García-del-Amo, André B. Junqueira, Vanesse Labeyrie, Xiaoyue Li, et al., "Local Indicators of Climate Change Impacts Described by Indigenous Peoples and Local Communities: Study Protocol," *PLOS One* 18, no. 1 (2023): e0279847.

Chapter 4

1. Jeremy Wilks, "As Climate Change Alters Our Weather Systems, How Can the Agricultural Industry Adapt to a Warming World?," *euronews.green*, January 7, 2022, https://www.euronews.com/green/2022/07/01/agriculture-vs-climate-change-can-we-feed-the-world-with-a-warmer-planet.
2. Joshua M. Rapp, David A. Lutz, Ryan D. Huish, Boris Dufour, Selena Ahmed, Toni Lyn Morelli, and Kristina A. Stinson, "Finding the Sweet Spot: Shifting Optimal Climate for Maple Syrup Production in North America," *Forest Ecology and Management* 448 (2019): 187–197.
3. Hirra Azmat, "Warmer Winters in the Kashmir Valley Are Leading to Early Flowering of the Gul Toor," *Mongabay*, January 25, 2023, https://india.mongabay.com/2023/01/warmer-winters-in-the-kashmir-valley-are-leading-to-early-flowering-of-the-gul-toor.

4. Saskia De Melker, "A Sour Season for Michigan's Cherry Farmers," *PBS News Hour*, August 16, 2012, https://www.pbs.org/newshour/science/science-july-dec12-michigancherry_08-15.

5. Joanna Adhem, "Spain Is amongst the World's Biggest Exporters of Almonds, so a Bad Harvest Can Have Serious Economic Implications," *euronews.green*, February 14, 2023, https://www.euronews.com/green/2023/02/14/flowering-into-frost-climate-change-is-destroying-spains-treasured-almonds; "Spain: Damage from Frost that Wipes Out 80% of the Fruit and Almond Harvest," PortalFruticola.com, April 12, 2022, https://www.portalfruticola.com/noticias/2022/04/12/danos-por-las-heladas-que-acaban-con-el-80-de-la-cosecha-de-frutas-y-almendras.

6. Susan M. Collins-Smith, "Late Freeze Causes Severe Damage to MS Blueberries," Mississippi State Extension Service, March 27, 2023, http://extension.msstate.edu/news/feature-story/2023/late-freeze-causes-severe-damage-ms-blueberries.

7. Andrew J. Allstadt, Stephen J. Vavrus, Patricia J. Heglund, Anna M. Pidgeon, Wayne E. Thogmartin, and Volker C. Radeloff, "Spring Plant Phenology and False Springs in the Conterminous US during the 21st Century," *Environmental Research Letters* 10, no. 10 (2015): 104008.

8. Lauren E. Parker and John T. Abatzoglou, "Warming Winters Reduce Chill Accumulation for Peach Production in the Southeastern United States," *Climate* 7, no. 8 (2019): 94.

9. "Bloom Watch," National Park Service, March 23, 2023, https://www.nps.gov/subjects/cherryblossom/bloom-watch.htm.

10. "The Economics of Fall Foliage Tourism in North Carolina," Appalachian State University Department of Biology, https://biology.appstate.edu/fall-colors/economics-fall-foliage-tourism-north-carolina.

11. William R. L. Anderegg, John T. Abatzoglou, Leander D. L. Anderegg, Leonard Bielory, Patrick L. Kinney, and Lewis Ziska, "Anthropogenic Climate Change Is Worsening North American Pollen Seasons," *Proceedings of the National Academy of Sciences* 118, no. 7 (2021): e2013284118.

12. "Climate and Health Outlook," Office of Climate Change and Health Equity, March 2023, https://www.hhs.gov/sites/default/files/climate-health-outlook-march-2023.pdf.

13. Evelyn F. Alecrim, Risa D. Sargent, and Jessica R. K. Forrest, "Higher-Latitude Spring-Flowering Herbs Advance Their Phenology More Than Trees with Warming Temperatures," *Journal of Ecology* 111, no. 1 (2023): 156–169.

14. J. Mason Heberling, Caitlin McDonough MacKenzie, Jason D. Fridley, Susan Kalisz, and Richard B. Primack, "Phenological Mismatch with Trees

Reduces Wildflower Carbon Budgets," *Ecology Letters* 22, no. 4 (2019): 616–623.
15. Anthony L. Westerling, Hugo G. Hidalgo, Daniel R. Cayan, and Thomas W. Swetnam, "Warming and Earlier Spring Increase Western US Forest Wildfire Activity," *Science* 313, no. 5789 (2006): 940–943.
16. Matthew W. Jones, John T. Abatzoglou, Sander Veraverbeke, Niels Andela, Gitta Lasslop, Matthias Forkel, Adam J. P Smith, et al., "Global and Regional Trends and Drivers of Fire under Climate Change," *Reviews of Geophysics* 60, no. 3 (2022): e2020RG000726.
17. Kristina A. Dahl, John T. Abatzoglou, Carly A. Phillips, J. Pablo Ortiz-Partida, Rachel Licker, L. Delta Merner, and Brenda Ekwurzel, "Quantifying the Contribution of Major Carbon Producers to Increases in Vapor Pressure Deficit and Burned Area in Western US and Southwestern Canadian Forests," *Environmental Research Letters* 18, no. 6 (May 2023), https://doi.org/10.1088/1748-9326/acbce.
18. "Wildfires and Climate Change," Center for Climate and Energy Solutions, accessed March 14, 2024, https://www.c2es.org/content/wildfires-and-climate-change.
19. US Forest Service, "The Rising Cost of Wildfire Operations: Effects on the Forest Service's Non-Fire Work," August 4, 2015, https://www.fs.usda.gov/sites/default/files/2015-Fire-Budget-Report.pdf.
20. Nicole E. Rafferty and Anthony R. Ives, "Effects of Experimental Shifts in Flowering Phenology on Plant–Pollinator Interactions," *Ecology Letters* 14, no. 1 (2011): 69–74.
21. Gabriel Reygondeau, Juan Carlos Molinero, Steve Coombs, Brian R. MacKenzie, and Delphine Bonnet, "Progressive Changes in the Western English Channel Foster a Reorganization in the Plankton Food Web," *Progress in Oceanography* 137 (2015): 524–532.
22. Eric Post and Mads C. Forchhammer, "Climate Change Reduces Reproductive Success of an Arctic Herbivore through Trophic Mismatch," *Philosophical Transactions of the Royal Society B: Biological Sciences* 363, no. 1501 (2008): 2367–2373.
23. Samantha Harrington, "Five Winners and Losers of Premature Springs," *Yale Climate Connections*, March 20, 2019, https://yaleclimateconnections.org/2019/03/five-winners-and-losers-of-premature-springs.
24. Samantha Chisholm Hatfield, Elizabeth Marino, Kyle Powys Whyte, Kathie D. Dello, and Philip W. Mote, "Indian Time: Time, Seasonality, and Culture in Traditional Ecological Knowledge of Climate Change," *Ecological Processes* 7, no. 1 (2018): 1–11.

25. Hatfield et al., "Indian Time," 5.
26. Post and Forchhammer, "Climate Change Reduces Reproductive Success," 2373; Karen H. Beard, Katharine C. Kelsey, A. Joshua Leffler, and Jeffrey M. Welker, "The Missing Angle: Ecosystem Consequences of Phenological Mismatch," *Trends in Ecology and Evolution* 34, no. 10 (2019): 885–888.

Chapter 5

1. A. K. Ettinger, C. J. Chamberlain, I. Morales-Castilla, D. M. Buonaiuto, D. F. B. Flynn, T. Savas, J. A. Samaha, and E. M. Wolkovich, "Winter Temperatures Predominate in Spring Phenological Responses to Warming," *Nature Climate Change* 10, no. 12 (2020): 1137–1142.
2. Theresa M. Crimmins, Michael A. Crimmins, and C. David Bertelsen, "Onset of Summer Flowering in a 'Sky Island' Is Driven by Monsoon Moisture," *New Phytologist* 191, no. 2 (2011): 468–479.
3. Nancy J. Turner and Andrea J. Reid, "'When the Wild Roses Bloom': Indigenous Knowledge and Environmental Change in Northwestern North America," *GeoHealth* 6, no. 11 (2022): e2022GH000612.
4. Ryan Haines, "Use of Traditional Phenological Knowledge Indicators to Predict Lake Sturgeon Spawning Timing on the Seine River," 2017, https://legacyfiles.ijc.org/tinymce/uploaded/36b_Seine_River_Sturgeon_Indicators_Report_for_Publication_1.pdf.
5. Trevor C. Lantz and Nancy J. Turner, "Traditional Phenological Knowledge of Aboriginal Peoples in British Columbia," *Journal of Ethnobiology* 23, no. 2 (2003): 263–286.
6. Richard J. Hopp and Helmut Lieth, "Introduction," in *Phenology and Seasonality Modeling*, ed. Helmut Lieth (New York: Springer-Verlag, 1974), 367–368.
7. Daniel A. Herms, "Using Degree-Days and Plant Phenology to Predict Pest Activity," in *IPM (Integrated Pest Management) of Midwest Landscapes* (Saint Paul: Minnesota Agricultural Experiment Station Publication, 2004), 58:49–59.
8. René A. F. de Réaumur, "Observation du thermometer, faites à Paris pendant l'année 1735, compares avec celles qui ont été faites sous la ligne, à l'Isle de France, à Alger et en quelques-unes de nos isles de l'Amérique," in *Mémoires de l'Académie des Sciences de Paris* (1735): 545–576.
9. Jen Yu Wang, "A Critique of the Heat Unit Approach to Plant Response Studies," *Ecology* 41, no. 4 (1960): 785–790.
10. "Corn Growth Stages and GDU Requirements," Agronomic Spotlight, accessed March 17, 2024, https://www.corn-states.com/app/uploads/2018/07/corn-growth-stages-and-gdu-requirements.pdf.

11. Kathleen S. Knight, Britton P. Flash, Rachel H. Kappler, Joel A. Throckmorton, Bernadette Grafton, and Charles E. Flower, "Monitoring Ash (Fraxinus spp.) Decline and Emerald Ash Borer (Agrilus Planipennis) Symptoms in Infested Areas," US Forest Service General Technical Report NRS-139, 2014, accessed September 16, 2023, https://www.fs.usda.gov/nrs/pubs/gtr/gtr_nrs139.pdf.
12. "Status of Spring," USA National Phenology Network, May 30, 2023, https://www.usanpn.org/news/spring.
13. Carlos M. Carrillo, Toby R. Ault, and Daniel S. Wilks, "Spring Onset Predictability in the North American Multimodel Ensemble," *Journal of Geophysical Research: Atmospheres* 123, no. 11 (2018): 5913–5926.
14. Ha Kyung Lee, So Jeong Lee, Min Kyung Kim, and Sang Don Lee, "Prediction of Plant Phenological Shift under Climate Change in South Korea," *Sustainability* 12, no. 21 (2020): 9276; Jung Gun Cho, Sunil Kumar, Seung Heui Kim, Jeom-Hwa Han, Catherine S. Durso, and Patrick H. Martin, "Apple Phenology Occurs Earlier across South Korea with Higher Temperatures and Increased Precipitation," *International Journal of Biometeorology* 65 (2021): 265–276.
15. Romualdas Juknys, Arvydas Kanapickas, Irma Šveikauskaitė, and Gintarė Sujetovienė, "Response of Deciduous Trees Spring Phenology to Recent and Projected Climate Change in Central Lithuania," *International Journal of Biometeorology* 60 (2016): 1589–1602; Zhi Hu, Huanjiong Wang, Junhu Dai, Quansheng Ge, and Shaozhi Lin, "Stronger Spring Phenological Advance in Future Warming Scenarios for Temperate Species with a Lower Chilling Sensitivity," *Frontiers in Plant Science* 13 (2022): 830573.
16. Scott N. Zimmer, Matthew C. Reeves, Joseph R. St. Peter, and Brice B. Hanberry, "Earlier Green-Up and Senescence of Temperate United States Rangelands under Future Climate," *Modeling Earth Systems and Environment* 8, no. 4 (2022): 5389–5405.
17. Janet S. Prevéy, Lauren E. Parker, Constance A. Harrington, Clayton T. Lamb, and Michael F. Proctor, "Climate Change Shifts in Habitat Suitability and Phenology of Huckleberry (*Vaccinium membranaceum*)," *Agricultural and Forest Meteorology* 280 (2020): 107803; Janet S. Prevéy, Lauren E. Parker, and Constance A. Harrington, "Projected Impacts of Climate Change on the Range and Phenology of Three Culturally-Important Shrub Species," *PLOS One* 15, no. 5 (2020): e0232537.
18. Peng Li, Zelin Liu, Xiaolu Zhou, Binggeng Xie, Zhongwu Li, Yunpeng Luo, Qiuan Zhu, and Changhui Peng, "Combined Control of Multiple Extreme Climate Stressors on Autumn Vegetation Phenology on the Tibetan Plateau

under Past and Future Climate Change," *Agricultural and Forest Meteorology* 308 (2021): 108571; Zimmer et al., "Earlier Green-Up and Senescence of Temperate United States Rangelands under Future Climate."

Chapter 6

1. Kate Yoder, "It's Not Just You: Everyone Is Googling 'Climate Anxiety,'" *Grist*, October 5, 2021, https://www.salon.com/2021/10/05/its-not-just-you-everyone-is-googling-climate-anxiety_partner.
2. Timothy Morton, *Hyperobjects: Philosophy and Ecology after the End of the World* (Minneapolis: University of Minnesota Press, 2014).
3. Ellen G. Denny, Katharine L. Gerst, Abraham J. Miller-Rushing, Geraldine L. Tierney, Theresa M. Crimmins, Carolyn A. F. Enquist, Patricia Guertin, et al., "Standardized Phenology Monitoring Methods to Track Plant and Animal Activity for Science and Resource Management Applications," *International Journal of Biometeorology* 58, no. 4 (2014): 591–601.
4. Erynn Maynard-Bean, Margot Kaye, Tyler Wagner, and Eric P. Burkhart, "Citizen Scientists Record Novel Leaf Phenology of Invasive Shrubs in Eastern US Forests," *Biological Invasions* 22, no. 11 (2020): 3325–3337.
5. Nathan Emery, Keely Roth, and Alexandria Lynn Pivovaroff, "Flowering Phenology Indicates Plant Flammability in a Dominant Shrub Species," *Ecological Indicators* 109 (2020): 105745.
6. Vijay V. Barve, Laura Brenskelle, Daijiang Li, Brian J. Stucky, Narayani V. Barve, Maggie M. Hantak, Bryan S. McLean, et al., "Methods for Broad-Scale Plant Phenology Assessments Using Citizen Scientists' Photographs," *Applications in Plant Sciences* 8, no. 1 (2020): e11315.
7. Lucas Rodriguez Forti, Fábio Hepp, Juliana Macedo de Souza, Airan Protazio, and Judit K. Szabo, "Climate Drives Anuran Breeding Phenology in a Continental Perspective as Revealed by Citizen-Collected Data," *Diversity and Distributions* 28, no. 10 (2022): 2094–2109.
8. Allen H. Hurlbert and Zhongfei Liang, "Spatiotemporal Variation in Avian Migration Phenology: Citizen Science Reveals Effects of Climate Change," *PLOS One* 7, no. 2 (2012): e31662.
9. Breanna F. Powers, Jason M. Winiarski, Juan M. Requena-Mullor, and Julie A. Heath, "Intra-Specific Variation in Migration Phenology of American Kestrels (*Falco sparverius*) in Response to Spring Temperatures," *Ibis* 163, no. 4 (2021): 1448–1456.
10. T. H. Sparks and P. D. Carey, "The Responses of Species to Climate over Two Centuries: An Analysis of the Marsham Phenological Record, 1736–1947," *Journal of Ecology* (1995): 321–329.

11. Ulf Büntgen, Alma Piermattei, Paul J. Krusic, Jan Esper, Tim Sparks, and Alan Crivellaro, "Plants in the UK Flower a Month Earlier under Recent Warming," *Proceedings of the Royal Society B* 289, no. 1968 (2022): 20212456.
12. Elisabeth Beaubien and Andreas Hamann, "Spring Flowering Response to Climate Change between 1936 and 2006 in Alberta, Canada," *BioScience* 61, no. 7 (2011): 514–524.
13. M. Bison, Nigel Gilles Yoccoz, B. Z. Carlson, and Anne Delestrade, "Comparison of Budburst Phenology Trends and Precision among Participants in a Citizen Science Program," *International Journal of Biometeorology* 63, no. 1 (2019): 61–72.
14. Michael J. O. Pocock, Iain Hamlin, Jennifer Christelow, Holli-Anne Passmore, and Miles Richardson, "The Benefits of Citizen Science and Nature-Noticing Activities for Well-Being, Nature Connectedness and Pro-Nature Conservation Behaviours," *People and Nature* 5, no. 2 (2023): 591–606.
15. Craig R. Williams, Sophie M. Burnell, Michelle Rogers, Emily J. Flies, and Katherine L. Baldock, "Nature-Based Citizen Science as a Mechanism to Improve Human Health in Urban Areas," *International Journal of Environmental Research and Public Health* 19, no. 1 (2021): 68.
16. Theresa M. Crimmins, Erin Posthumus, Sara Schaffer, and Kathleen L. Prudic, "COVID-19 Impacts on Participation in Large Scale Biodiversity-Themed Community Science Projects in the United States," *Biological Conservation* 256 (2021): 109017.
17. Freeman Tilden, *Interpreting Our Heritage: Principles and Practices for Visitor Services in Parks, Museums, and Historic Places* (Chapel Hill: University of North Carolina Press, 1957), 63.

FURTHER READING

Cavalier, Darlene, Catherine Hoffman, and Caren Cooper. *The Field Guide to Citizen Science: How You Can Contribute to Scientific Research and Make a Difference*. Portland, OR: Timber Press, 2020.

Chisholm Hatfield, Samantha, Elizabeth Marino, Kyle Powys Whyte, Kathie D. Dello, and Philip W. Mote. "Indian Time: Time, Seasonality, and Culture in Traditional Ecological Knowledge of Climate Change." *Ecological Processes* 7, no. 1 (2018): 1–11.

Cohen, Jeremy M., Marc J. Lajeunesse, and Jason R. Rohr. "A Global Synthesis of Animal Phenological Responses to Climate Change." *Nature Climate Change* 8, no. 3 (2018): 224–228.

Emanuel, Kerry. *What We Know about Climate Change*. Cambridge, MA: MIT Press, 2018.

Hineline, Mark L. *Ground Truth: A Guide to Tracking Climate Change at Home*. Chicago: University of Chicago Press, 2018.

Lepczyk, Chris, Timothy Vargo, and Owen D. Boyle, eds. *Handbook of Citizen Science in Ecology and Conservation*. Oakland: University of California Press, 2021.

Lieth, Helmut, ed. *Phenology and Seasonality Modeling*. Vol. 8. New York: Springer Science and Business Media, 1974.

Parmesan, Camille, and Gary Yohe. "A Globally Coherent Fingerprint of Climate Change Impacts across Natural Systems." *Nature* 421, no. 6918 (2003): 37–42.

Pörtner, H.-O., D. C. Roberts, E. S. Poloczanska, K. Mintenbeck, M. Tignor, A. Alegría, M. Craig, et al., eds. 2022. *Climate Change 2022: Impacts, Adaptation, and Vulnerability: Contribution of Working Group II to the Sixth Assessment Report of the Intergovernmental Panel on Climate Change*. Geneva, Switzerland: IPCC.

Primack, Richard B. *Walden Warming: Climate Change Comes to Thoreau's Woods*. Chicago: University of Chicago Press, 2020.

Schwartz, Mark Donald, ed. *Phenology: An Integrative Environmental Science*. Vol. 132. Dordrecht, South Holland: Kluwer Academic Publishers, 2003.

Turner, Nancy. *Ancient Pathways, Ancestral Knowledge: Ethnobotany and Ecological Wisdom of Indigenous Peoples of Northwestern North America*. Vol. 74. Montreal: McGill-Queen's University Press, 2014.

INDEX

THERESA CRIMMINS is an associate professor in the School of Natural Resources and the Environment at the University of Arizona and director of the USA National Phenology Network. She works enthusiastically to cultivate a broader appreciation of phenology among scientists and nonscientists alike. Theresa has published over seventy peer-reviewed articles and book chapters; her writing has also appeared in *Scientific American*, the *Hill*, the *Old Farmer's Almanac* and many other outlets.